Locksmithing, Lock Picking & Lock Opening
Professional Training Manual

WILLIAM PICKET

First published in Great Britain by Swordworks Books

ISBN 978-1906512439

Printed and bound in the UK & US

A catalogue record of this book is available from the British Library

Cover design by Swordworks Books

FOREWORD

Historically locksmiths used to make the entire lock, working for hours hand cutting screws and doing much file-work. Today, the rise of mass production means that this is no longer true and, though a few expert locksmiths are also engineers and capable of sophisticated repairs and renovation work, the vast majority of locks are repaired by swapping of parts or like-for-like replacement - or upgraded to modern mass-production items.

Safes and strongboxes are the exception to this, and to this day large vaults are custom designed and built at great cost.

Although fitting of keys to replace lost keys to vehicles and homes and the changing of keys for homes and businesses to maintain security are still an important part of locksmithing, locksmiths today are primarily involved in the installation of higher quality lock-sets and the design, implementation and management of keying and key control systems.

In terms of physical security, a locksmith's work frequently involves determining the level of risk to an individual or institution then recommending and implementing appropriate combinations of equipment and policies to create security layers which exceed the reasonable gain to an intruder or attacker. The more different security layers are implemented the more additional skills and knowledge and tools to defeat them are required. Because each layer is an expense to the customer the application of appropriate levels, without exceeding reasonable costs to the customer, is often very important and requires a skilled and knowledgeable locksmith to determine.

For the beginning locksmith, locksmithing may offer the route to a secure job or business, being able to offer a service without the necessity of large premises. In fact, a vehicle and a selection of tools and spare locks and parts is all that is required. Apart, that is, from a degree of skill.

This book is aimed at achieving this kind of skill and operation.

William Pickett

TABLE OF CONTENTS

INTRODUCTION

A locksmith is someone who specialises in locks. Anyone who has been locked out of their car or home is probably familiar with one of the services a locksmith offers, but locksmiths do a lot more than helping people get back into their locked homes. This is an old profession. As long as there have been locks, and keys to fit them, there have been people to specialise in them. In addition to handling locks some locksmiths also work in security consulting, since locks are one way to control access to something.

As with any profession with "smith" in the title, locksmiths originally fabricated locks and keys from metal, although modern locksmiths also work with electronic locks. Locksmiths install, repair and adjust locks in everything from cars to office buildings. They also offer services to people who are locked out or individuals who want to consult with someone about their security systems.

Most locksmiths need a lot of equipment for the practice of their trade and it is not uncommon for a locksmith to use a van to carry their tools around. Locksmiths typically carry tools which are designed for picking locks, so that they can get into locked structures and vehicles. They have equipment to fabricate keys and diagnose problems with a wide variety of locks and security systems. When a locksmith installs locks on an office building they usually become that building's default locksmith, because of their presumed familiarity with the building's locks. Some locksmiths retain extra key sets, by request, in order to gain quick entry when needed. They also keep records of the types of locks used in a structure, replacement or repair work undertaken and any other relevant information. This can be such as connection with a security system which demands that a code be entered when a door is unlocked.

To become a locksmith some people train as apprentices. It helps to have metal fabricating, construction and electronics skills, but these are not essential. Locksmiths perform such specific tasks that these skills can be learned on the job. With the shortage of apprenticeships and more conventional routes to training, many people are having to adopt the home study approach to learning their new trade. Good locksmiths are excellent problem solvers who are willing to work at unusual hours, and they are of course discreet, as they often handle confidential or sensitive information.

A Master Locksmith is certified by the Master Locksmith Association (MLA). The accreditation process involves a four–part exam covering written and practical aspects. However, before you become a Master, you must learn the ropes. For a locksmith this consists of three exams and qualifications offered by the British Locksmiths Institute (BLI), which is a part of the MLA.

The MLA is the Locksmiths' trade body and though they have no regulatory role, they are recognised by the police, the Home Office, the British Standards Institute and the Association of British Insurers. Their qualifications are also recognised throughout the industry.

Locksmiths may be:-

- commercial (working from premises)
- mobile (working from of a vehicle)
- institutional (employed by an institution
- investigational (forensic locksmiths)
- a specialist in one aspect of the skill (such as an auto lock specialist, a master key system specialist or a safe technician)

Many (not all) are also security consultants, but not every security consultant has the skills and knowledge of a locksmith. Locksmiths are frequently certified in specific skill areas or to a level of skill within the trade. This is separate from certificates of completion of training courses. In determining skill levels, certifications from manufacturers or locksmith associations are usually more valid criteria than certificates of completion. Some locksmiths decide to call themselves "Master Locksmiths" whether they are fully trained or not and some training certificates appear quite authoritative.

In this book, we aim to introduce the reader to the techniques of locksmithing as they apply to lock opening, commonly called lockouts, where the property owner has lost their keys or the lock has been damaged. In this way the trainee will gain valuable experience of the working of the inner mechanisms of locks. We also cover fitting of entire lock mechanisms in its most basic manner, although this kind of work often moves into the realms of carpentry and would best be covered elsewhere.

As with all skills the rule is:-

- practice
- practice
- practice

If you work through this manual and keep practicing on every different type of lock available, you will quickly acquire the necessary skills to enable you to call yourself a locksmith.

A SHORT HISTORY OF LOCKS AND LOCKSMITHING

Locksmiths are professionals who design and service locks and other hardware mechanisms. The term locksmith refers to an individual who shapes metal into locks. Though locksmiths rarely handcraft locks today, the name is still used to refer to any hardware-related professional.

Locksmithing is one of the oldest handicrafts known to civilised man. Long before the great Pyramids were built, locksmiths plied their trade in Egypt, Babylon, Assyria and China. In fact, it may be said that the first key to be used by mankind was the branch of the tree which the caveman used to move aside the boulder that guarded the entrance to his cave. In the ruins of ancient cities, archaeologists frequently uncover locking devices that protec¬ted the wealth of men who lived before the time of written history.

Over forty centuries ago an Egyptian artist painted a fresco on an ancient temple which showed a lock that was then in use. A similar lock was actually found in the ruins of a once sumptuous palace in a suburb of the biblical city of Nineveh. This lock is said to be the oldest lock in existence.

The pattern, however, was widely imitated and even to this day, similar locks are occasionally dug up in places as far remote from ancient Assyria as Scotland, Japan, and even America. Apparently the skill and techniques of the ancient lock makers survived the fall of great empires and even time it¬self because the basic mechanical principle of the so-called Egyptian lock is still being used in our modern pin tumbler locking devices. The lock men¬tioned in the Bible as being used on the House of David is, in effect, the same type that is used today on modern house doors.

The Egyptian type lock was created only for men of great wealth. Later models were made of brass and iron. They were ornamented with in¬laid pearl, gold and silver. The poor man still relied on the wooden cross bar to keep his home safe against outside attack. It is said that the ancient Greeks were the first to do something about a key operated device for the lower class¬es. In the days of Homer they used to tie their doors shut with intricately knotted ropes. They were so cleverly tied that only the owner could find the correct method of unknotting them. Under the superstition of the times no one would dare tamper with the ropes, lest a curse fall upon them and their families.

But the robbers' greed overcame superstition and the Greeks were even¬tually forced to discard the knotted rope lock in favour of a more substantial locking device. This arrangement consisted of the usual cross bar on the in¬side, and a large hook shaped key the size of a farmer's sickle which could be inserted through a hole in the door to push the bar aside. This lock was a very effective barrier against the burglars of that time. Also the huge key came to be used as an excellent weapon of defence.

The ancient Romans brought locksmithing to new heights of achievement. They combined the Egyptian and Greek features and produced excellent me¬chanical door locks that were installed on the inside of the door and operated with a key from the outside through a key hole.

Although the Chinese and Near East peoples developed the padlock, the Romans are credited with having popularised it through the world of their day. They made it a practical device, even going so far as to make the key part of a finger ring for convenience in carrying. The fact that the Romans had no pockets in their togas inspired their locksmiths to devise keys that were small and inconspicuous. The present trend of twentieth century lock makers is in the same direction.

Every schoolboy learns about the ruins of the ancient city of Pompeii which was engulfed by the volcano Vesuvius in 64 A. D. When archaeologists were unearthing the buried city they came across a house that was evidently the site of a locksmith shop. They found many types of door locks, padlocks and highly ornamented keys, some of which were inlaid with silver and gold. Skewers and odd shaped prongs were also discovered. Undoubtedly, these were the picks that were used by this ancient craftsman to unlock his client's doors or padlocks in an emergency.

The origin of the warded lock is obscure. In the dim past some ingenious lock maker discovered that he could place a series of obstructions in the path of a turning key and thus make the lock secure. Only the correct key which had corresponding spaces on its body could bypass the obstructions. Some historians credit the Etruscans in northern Italy with this invention. But evi¬dence has recently come to light to prove that this mechanism was known to the Greeks and Romans, too. Regardless of its beginning, this type of lock became the most widely used from later Roman times up to the time when the pin tumbler cylinder lock was invented.

After the Roman Empire had declined, the world entered into the period known as the Middle Ages. Although this was generally a period when science and education suffered severe neglect, locksmithing seemed to flourish. Ward¬ed locks were not only mechanical devices, but also works of great artistic creation. These were the days of medieval castles and knights. Robber barons employed locksmiths to create security devices which were both secure and rich in design. It was a point of pride and prestige to have the handsomest locks guarding the treasures of the castle. Also at this time the great monasteries developed. In them reposed all forgotten learning, the books and manuscripts of the ancient days. Locksmiths turned their skill toward creating the locks that guarded man's knowledge until the world was ready to absorb it once again.

The traditions and customs of locksmithing as it is known today are de¬rived from the great medieval guilds of locksmiths which had their formation during the Middle Ages. The Guild was the all-powerful force. It regulated the terms of apprenticeship, the rules and conduct for journeymen and the techniques of the masters. It regulated everything from prices to the number of rivets that could be placed in a lock. It was supreme master over the craft, and the penalty for defiance was expulsion and the loss of the right to earn a living as a locksmith. In the earlier days the strictness of the Guild Masters helped maintain the integrity of the locksmith's trade. But like all autocratic and dictatorial organisations it became overbearing, jealous of its power and unwilling to advance. The trade became a father-to-son enter¬prise and followed the guild pattern right up to the nineteenth century. Con¬sequently, lock design did not progress much beyond the creation of intricate designs to bewilder thieves. False keyholes, false wards and lots of "ginger¬bread" design were heaped upon the lock. Chests that cut off fingers, fired pistol shots or ejected murderous knives were part of the locksmith's stock in trade. Secret panels and hidden locations for the locks foiled the would-be burglar because he could not find where to begin his lock opening attempt.   However, the basic structure of locks remained the same as in the days of the Romans.

While the rest of the world awakened to the wonders of science, art and engineering during the next period of history, the Renaissance, locksmiths still amused themselves and their customers with trick locks and fancy de¬signs. Only one notable improvement was added to locks during this period. The lever tumbler was introduced. At first the lever was used merely to hold the lock bolt in position so that it would not move either backward or for¬ward, unless the key lifted the lever out of the way. This basically sound prin¬ciple was never fully appreciated by Renaissance locksmiths who handed it down to their more skilful heirs of the 19th century.

Up to this point lock picking was comparatively simple. Once a thief could get past the trick devices he could easily retract the bolt with simple tools. The world sorely needed a lock that would foil the burglar even though he knew its construction thoroughly.

While it is true that combination locks (sometimes called letter or num¬ber locks) made their appearance during various stages of the world's history, it must be noted that these locks were usually confined to padlock construc¬tion. They offered very little security against opening by feel and were never popularly accepted as secure locking devices until modern machine methods made them works of precision. In many cases numbers were used instead of letters and sometimes only symbols.

However most combination padlocks used letters. When turned to form the correct word, the padlock would open. The advantage of locks without keys was purely of convenience. Security was sacrificed. Combination padlocks were often looked upon as items of play rather than serious devices to protect property. It was often considered great sport to spend an evening at the fireside attempting to guess the correct combination of a new lock.

In 1778 Robert Barron of England invented a lever tumbler lock that of¬fered the first real security against picking. Instead of using a lever merely to support the bolt against movement, he used the lever to actually imprison or block the bolt and thus prevent its movement until the correct key was used. By using several lever tumblers, instead of one, he was able to con¬struct a lock that was fundamentally sound. For the first time since the al¬most forgotten Egyptian lock, the locking action was applied to the lock bolt itself. Gone forever was the need of tricks and false de-signs. True lock construction had emerged at last.

Notable inventors followed Barron's lead, and locksmithing began a new page in its long history. Names such as Bramah, Newell, Chubb Andrews, and Pettit are recorded in connection with improved design and in¬creasing security. These men were scientists who challenged the world and each other to pick their locks. Public contests were held and the masters pitted their lock picking skills against one another. The intense rivalry led to the greatest improvement in locks in over three thousand years. The most famous of these contests was held at the London Exhibition in 1851 when A. C. Hobbs, an American locksmith, picked open the best locks that the English had produced.

In the mid 1700s locks were few in the Colonies and most were copies of European mechanisms. With the founding of the Republic and the new prosperity there was a growing demand for sturdy door locks, padlocks and locks for safes and vaults. The American lock industry had its start. Each native craftsman had his own ideas about security. Between 1774 and 1920 American lockmakers patented some 3000 varieties of lock devices. Among them was the patent for a domestic lock by Linus Yale, Sr. This lock was a modification of an old Egyptian pin-tumbler principle that used a revolving cylinder.

In the early 1920s Walter Schlage advanced the concept of a cylindrical pin-tumbler lock by placing a push-button locking mechanism between the two knobs. Emphasis was on security, yet equally important to the modern architect and decorator, the lock became an intricate part of the door design. It was now possible to select complimentary styles of locks, metals and finishes. The revolutionary Schlage lock is a completely different concept of a cylindrical lock with the button-in-the-knob mechanism placed between the knobs, introduced by Walter Schlage in the early 1920s.

In America, locksmithing was not fettered by old guild traditions and customs. Locksmiths took full advantage of the advanced mechanical know¬ledge of the day and produced masterpieces of precision and security. The two outstanding names of 19th century American lock history are Yale and Sargent.

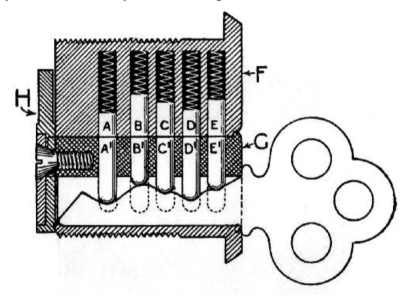

Linus Yale, Sr was a successful bank lock maker. However, his greatest claim to fame is his invention of the pin tumbler cylinder lock in 1844 incorporating the basic principle of the early Egyptian lock. Linus Yale, Junior, the son, following in his father's footsteps, produced bank locks and improved the design of the pin tumbler lock so that it could be made by mass production methods and still provided excellent security. It was he who invented the grooved type of cylinder key that bears his name.

James Sargent is generally associated with the development of time locks. Although the idea of the time lock had been dreamed of in the days of the guilds, Sargent was the first to make a practical specimen. As mass production methods were introduced the locksmithing craft broke into two distinct groups:-

- lock makers
- lock repairmen

The manu¬facturing end of the business became a highly technical and complex function, leading to the creation of giant manufacturing organisations.The term "locksmith" now refers to the repairman rather than the manufacturer of locks.

With the advent of the automobile, lock design took a new step forward. Entirely new constructions were demanded. Locks that could withstand exces¬sive vibration were developed. This phase of lock making began shortly after the First World War. It is continuing up to the present moment with new de¬signs and new ideas being constantly introduced. The locksmiths function has clearly become that of repairing or replacing these locks and of fitting keys to them.

In recent times, several remarkable inventions have increased the scope and efficiency of locksmiths. The first of these, the key duplicating machine, was invented by Henry Gussman in 1909. This machine ended the drudgery of hand filing and made accurate reproduction of keys within a matter of minutes a practical reality.

The next great advance was the development of the code machine in 1926 by the engineers of the Independent Lock Company. This creation opened up a new field for locksmiths because it enabled them to make keys to locks ac¬cording to their serial numbers.

Many inventions for making the art of locksmithing more scientific and more lucrative have been introduced in recent years. Locksmiths will be for¬ever indebted to men like H. Hoffman, Ted Johnstone, Eli Epstein and scores of others whose tools, devices and methods have become standard in the lock¬smith industry. These men inventing tools that enable locksmiths to measure lock tumblers without requiring the disassembly of the lock, and for effective devices to open locks in emergencies.

By the early 20th century, the locksmithing profession had changed dramatically. The Industrial Revolution allowed for mass production of locks, freeing up the locksmith who had traditionally hand-made all locks and components. This has allowed modern locksmiths to focus on security consultation, assisting the public with emergency lock-picking and other lock-related needs, and developing new technologies to improve the industry.

Locksmithing is a vital, growing industry. With modern techniques and scientific tools, it has progressed from a small hand craft to a flourishing service industry with a bright future in the years to come.

Tools and Equipment

A locksmith is no different to any other craftsmen and will need a variety of both general and specialised tools to carry out their trade. Some of these locksmith tools will need to be purchased as they are highly specialised precision tools that can only be obtained through trade sources. However, the locksmith does have one tremendous advantage in that some of the tools of their trade can be made themselves in their own workshop.

The first requirement and possibly the most important is a good workbench that will provide a working surface for the various jobs undertaken, together with storage for tools, equipment and spare locks and parts. If possible position the workbench where there is plenty of daylight as locks can be intricate to work on and the more daylight that is available the easier this task will be. If necessary, provide strong artificial light to supplement the daylight that is available. The bench should be a convenient height, usually about belt level. It should be provided with at least one professional vice and possibly a smaller jeweller's vice for fine work. Swivel vices are not ideal as they can allow a degree of movement which reduces the ability of the vice to hold the working piece firmly. Ideally you need a heavy machinist's vice with jaws that at least four inches across and which will open to allow a depth of at least 5 inches. Near to the vice you should have several trays and draws in which you can keep a variety of lock spares, pins and springs etc. that are always immediately to hand. Try to keep the bench clear and if possible have a back panel above the bench to hang your tools.

You will need to a selection of trays for jobs that you are working on, so that the parts do not stray from the lock they belong to, and also for completed jobs. Make sure that when a customer brings in a lock for repair you attach a tag to it so that the job does not get lost amidst a pile of similar work. On the tag you should make sure you record exactly what the customer requires, what kind of repair, what kind of finished job, how many keys they want, together with a record if applicable of the estimated price that you have given to the customer.

A key machine is virtually essential for the working locksmith. You should consider mounting this on a separate table or bench, as when you are working at your bench vibrations from

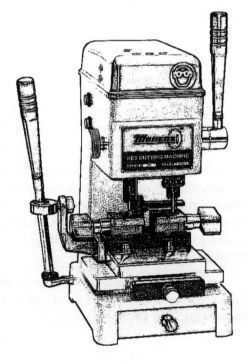

hammering and similar operations can jar a key machine out of adjustment. Even the tiniest variation in a key machine will obviously cause difficulties, in that the keys you produce will not be accurate and will not open the lock they are designed for.

When choosing a key machine make sure it is as versatile as possible, with a range of changeable vices and holders to allow for as many types of key as possible. The machine you choose should be capable of accurately duplicating Yale or cylinder type keys, flat steel keys and a variety of other common keys. Used machines are not always a good idea as they can be well worn and result in continual complaints from customers that keys do not work.

You will need a variety of bench tools. To begin with, you will need the following:-

- a pair of wire cutters
- good quality torch
- selection of high quality screwdrivers
- selection of jeweller's screwdrivers
- assortment of small files with handles
- 8 inch Mill file, 6 inch Mill file, 6 inch round file, 6 inch square file, 6 inch triangular file, 4 inch flat warding file, 6 inch flat warding file (or metric equivalent)
- good-quality electric drill with a sorted professional steel bits
- selection of counter sinks
- hammer
- tweezers
- key caliper
- steel rule
- half inch wood chisel
- graphite airgun
- a can of thin oil
- grease

A key caliper is required to accurately measure the depth of cuts in keys. Other than that, almost all of these tools are commonly available. The rule to keep in mind is to get as high a quality of tools as is possible within your budget. A graphite gun is the locksmith's friend and will have helped to free locks than old, worn and possibly

jammed by pumping powdered graphite into the mechanism. It is possible to use oil for this particular task but it tends to attract dust and grit and can clog the working of the lock.

You will need a tool known as a thimble. This is completely different to a sewing thimble. The thimble is used to hold the lock plug while it is being serviced.

Another necessity is a broken key extractor for pin tumbler locks. This can be made on the bench and generally consists of a short length of piano wire with a small hook formed at the end of the wire. The extractor is slipped into the keyhole of the lock alongside the broken key and hooks over the point of the key thus allowing it to be pulled out. A similar tool can be made with a piece of thin fret saw blade mounted in a handle.

A laying out board is most essential for working on and repairing pin tumbler cylinders. As you dismantle the cylinder you will be able to lay out the springs and small parts. Where you are working on master keyed locks this is even more essential as the slightest variation will mean that the master key will not fit all of the locks, which defeats the object of having a master key in the first place.

You should make the laying out board from a flat piece of plywood, cutting eight parallel V shaped grooves across the wood to correspond to the tumbler positions of the pin tumbler lock. The majority of locks may only have five or six pins, but in the case of higher security locks as used in commercial premises you may well need to accommodate as many as eight pins. At right angles to the grooves draw three lines across the groove, dividing each groove into four sections, one for the spring of the tumbler and one for each pin segment. It is then possible to distribute the tumblers piece by piece across-the-board as they are removed from the lock. In this way you should never mislay or misplace any part of the lock you are working on.

As your work develops you should equip yourself with additional equipment such as picks and master keys together with a range of different keys for opening large and small locks, ranging from small cases and bags to large commercial doors and other applications.

You will also need to equip yourself with a box of tools and equipment to take with you on callouts. These should include a selection of all of the small bench tools, files, hammers, special lock tools etc. and of course a range of carpenter's tools for as well as opening locks where the keys had been lost or jammed in the mechanism. You will also need to install and replace locks with new ones.

An example call out toolbox would be as follows (you may use metric equivalents):

- small screwdriver
- medium screwdriver
- large screwdriver

- half inch wood chisel
- 6 inch sheet metal snips
- 16 ounce hammer
- small hammer
- 6 inch side cutters
- 8 inch end cutting nippers
- 12 inch hacksaw
- keyhole hacksaw
- canvassing drill for metal
- rechargeable drill with set of high-quality steel bits
- assortment of small taps and dies
- tap wrench and die holder
- graphite gun
- assortment of files
- torch
- steel rule
- assortment of specialist lock opening and servicing tools
- expansion bit
- wood bits
- centre punch
- thread gauge
- magnet and string
- broken screw extractor
- small soldering iron
- hand vice
- wrench
- specialist lock wrenches as necessary
- broken key extractor

As you build up your business you should begin to build a stock of key blanks and small repair parts for the different types of locks that you come across. Clearly stocking all of the parts and blanks that are available on the market would be impossible, especially when you are beginning. Therefore obtain parts catalogues for as many different lock types as possible so that you are able to quickly identify and order spare parts as required. You should consider carrying a small stock of small new pin tumbler or Yale lock tight barrels and a selection of padlocks so that you are able to make replacements quickly and easily and charge the customer accordingly.

A useful tool for the travelling locksmith is the snap pick. Also known as lock pick gun, pick gun or electric lock pick it is an alternative to conventional lock picking methods, which use a tension wrench and a lock pick to open a lock. Snap guns make the task of opening a lock much simpler.

The first snap guns were developed decades ago in order to assist police officers who were incapable of picking a lock to do so with minimal instruction. Rather than opening locks by the traditional raking techniques, a snap gun uses a primary law of physics, the transfer of energy, to open locks. Operating on the same principle as that of a cue ball, the snap gun strikes all of the bottom pins at once sending the driver pins up into the lock. This only lasts for a fraction of a second because the springs will force the pins back down into the lock. The tension wrench is also required in this situation.

If used properly, snap guns can be faster than traditional lock picking, but their use may risk damage to the lock. There are two types of snap gun, the spring home made type or the more automatic type, both manual and electric.

Below is a homemade version of a mechanical pick gun constructed out of stiff wire. This tool is referred to as the Snap Pick.

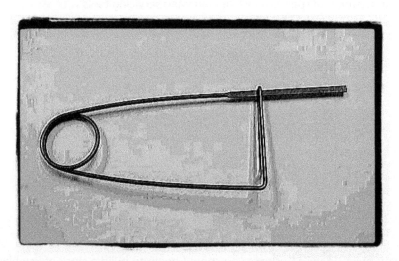

Here are some basic images showing how it is done.

The tip of the tool has been flattened to allow the blade to be manageable inside the keyway. The tool works by inserting the blade into the lock and pressing down on the spring coil, pictured below.

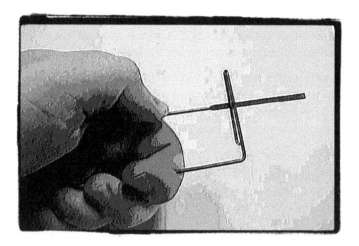

Once the coil has been compressed by pressing down on the wire, the wire is released. This action will snap the pins upward within the keyway simulating the action of a mechanical pick gun.

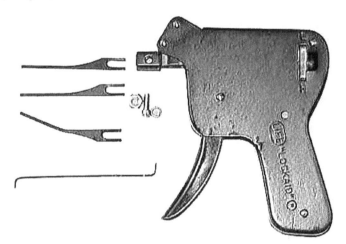

Lock pick guns work by creating a sharp impact within a pin tumbler lock. When these locks are at rest, top pins cross the shear line between the lock shell and the lock plug. The shear line needs to be clear before the lock will turn. When you put the proper key into the lock it pushes all the top pins into the shell, but keeps the bottom pins in the plug leaving the shear line clear. Wrong keys leave top and bottom pins crossing the shear line.

A pick gun bounces the pins. A hardened steel needle strikes the bottom pins sharply all at once. Top pins rest on bottom pins, and the impact is transferred to them just like a rear-ended car jars the car in front of it. Springs in the pin chambers push the top pins down at all times, but spring resistance is pretty weak. After the jarred pins bounce there is a moment when the shear line is clear, before the springs have pushed the top pins back to their resting positions. In this moment the lock is picked and can be turned.

Lock pick guns do not work on wafer tumbler locks which do not have top and bottom pins. A person may occasionally open one of these locks with a pick gun, but only when the pick needle is used to pick the lock in a conventional fashion. Pick guns destroy wafer locks easily, so do not use them on wafer locks.

Pin tumblers are primarily found in house locks and wafer tumblers are primarily found in automobiles. Pick guns can accomplish sometimes what conventional lock picks cannot. Some types of pick resistant pins (spool, mushroom, and serrated pins) defy conventional picking techniques, but can be bounced out of the way with a pick gun.

Using a pick gun requires precision, a delicate touch, excellent timing and practice. Practice is certain to destroy the locks you are working with, so be warned. Pick guns are NOT non-destructive tools.

Pin tumblers are made of brass and bottom pins are pointed. Very little wear occurs when a brass key is used in a lock but when the sharp edge of a pick gun's needle rams into the pin it leaves a dent. Often a pick gun must be triggered ten to fifteen times to open a lock and these dents compound on a very critical part of the bottom pin - the tip. When it becomes too deformed it will not rest properly on the key. It will be difficult to get the key in and out of the lock, and the lock may cease to function. Pick guns also break locks apart internally. They sometimes cause the pin retainer to fly off, letting springs shoot out of the pin chambers. This is not good and not easy for a non-professional to fix. Locking door knobs are especially susceptible to this.

Pick guns can be valid tools but most locksmiths go months at a time without using them, although they use their conventional picks every day.

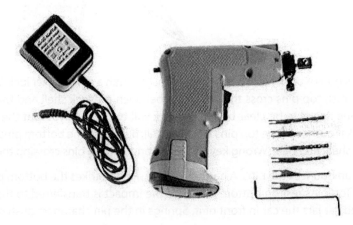

Turning tension is the key ingredient to success with a pick gun. It should be kept to the barest minimum, or else sideways binding pressure will prevent the top pins from bouncing. There are a few common and successful ways of applying tension while using a pick gun. The most successful for opening locks with pick-resistant top pins is one of pure timing.

Place the tension wrench in the lock, but do not apply any pressure until a split second after the pick needle has snapped against the bottom pins. This allows the top pins to shoot straight up into the shell before the plug is turned. Spool, mushroom and serrated pins all count on sideways binding pressure to hamper picking attempts. This method defeats them, but timing really has to be worked on. It is not easy.

Initial tension, kept just enough to prevent the tension wrench from falling out of the lock, is fine for most pin tumbler locks, especially for manually operated or homemade pick guns. Less tension is needed to use a pick gun than is necessary to rake a lock cylinder with standard picks. Beginners have a habit of cranking up their turning pressure as the pick snaps. This is counterproductive. Feathering is the most successful tension technique when using a pick gun. You need to vary the amount of tension as you snap the pick. Bound pins pushed too far have a chance to drop back into place and the odds of proper tension the moment after the pick snaps increase over a set push or a single nudge.

Pick guns are not always the answer. They do not work if the keyhole is curvy, the lock does not use pin tumblers or if the lock does not work smoothly. It takes practice to use them and the practice is hard on locks. If you are going to use one put as little pressure on the tension wrench as possible. This will allow you to use less snapping force with the gun, which leaves less damage inside the lock. Commercial pick guns generally have a dial on the side to set the snapping force. An emergency pick gun can be made from a coat hanger, a bobby pin and a piece of tape. It is not strong enough to pop the pin retainer off a lock shell's bible, although it does dent the bottom pins.

A tension wrench is necessary in conjunction with a pick gun. They are really easy to make. It is wise to use a tension wrench with a longer handle than normal to keep you aware of how much tension you are using. The tension wrenches that come with commercial pick guns are too short.

Pick guns do open locks, but they are not legal for most people to carry around. If you learn other ways to open locks without keys you will not often have the need for one. Since a simple one can be made without tools within a few minutes, it probably is not a necessary investment to order one. Electric pick guns cost more than the manual variety, and damage pins at a very rapid rate. If you have a pinning kit and the knowledge needed to replace damaged bottom pins, they can be good tools to use. However, battery operation does not remove the requirement for operator skill to successfully use them.

Another useful tool for the locksmith is the Slim Jim. In most cars a control arm extends from the inside end of the door lock cylinder. When the door key is turned the control arm and connecting linkage are activated to lock or unlock the door. The door unlocking tool is used when the door key is unavailable and only with the vehicle owner's knowledge and consent.

Before using this tool, consult the vehicle manufacturer's maintenance and parts manual to become familiar with that particular door locking mechanism. Door lock mechanisms vary from model year to year and from one manufacturer to another.

The control arm may be located to the front or rear of the door lock.

Always start with the tool pointing directly toward the lock. The tool is inserted on the outside of the window between the glass and weather-stripping. Move the tool along joint at window and probe for easy entry past the weather-stripping keeping tool in close contact to surface of glass. Move the tool to desired location above lock before penetrating further.

Now push tool down on the control arm. Some bend in the tool maybe required due to the thickness of the door or the distance between the glass and the control arm. Generally the lock can be activated to open by pushing down on the control arm. Occasionally the control arm must be pushed to the rear or lifted.

There is a slight bend near each end of the tool. When the tool is held flat against the glass, the bend will cause the end of the tool to project out and skid over the lower glass channel. The tool is generally used with the bent end pointing outward. The notched end will have to reach past the inside of the glass channel to lift the arm with the lock rod attached. The tool must have a bend to reach across the channel.

For the bent end to skid over the glass channel, the curve above it must be flattened out against the glass. If the tool is bent too much it can pass the end of the lock arm. If not bent enough, it will slide up in the glass channel making little if any sound.

Lock Opening Tools

As well as your basic tool kit, snap picks and slim jims, you will need lock picks for those occasions when you are called out to open a lock where the owner has lost their keys.

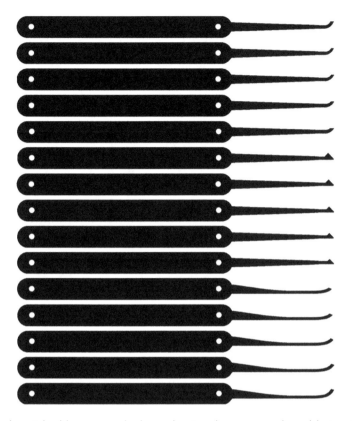

Locks can be picked because a lock mechanism becomes vulnerable to picking for two basic reasons:-

- design shortcomings and

- manufacturing shortcomings

Both of these flaws are directly related to the selling price of the locking device. The design flaw allows a pick, wire, pick key, paper clip, hair pin, knife blade, etc. to be inserted into the keyway in such a manner as to reach and operate the mechanism. Manufacturing shortcomings are found in loose tolerances in the manufacturing process. A tolerance is a necessary sloppiness that is found even in the most expensive of machine products. The closer to perfection the higher the cost will be. Therefore, whether in a lock or an automobile, a compromise must be arrived at the engineering level.

The vulnerability of tolerance is usually found in areas such as pin diameters versus pinhole diameters and a row of pin holes deviating from a straight line. Tolerances allow shims to be inserted in the small space necessary between moving parts. Tolerances allow a combination lock to reveal its inner secrets to a skilled manipulator. Tolerances are ever present in any machined product. They cannot be eliminated, only minimised which directly affects the cost of the part.

Simple warded locks are typically found in cheaper padlocks, file boxes, luggage, etc. The keys are usually stamped from flat steel and nickel plated. Higher quality warded locks sometimes use corrugated keys in an effort to provide better security and also make the key stronger. Each key has the ward cuts in a slightly different position.

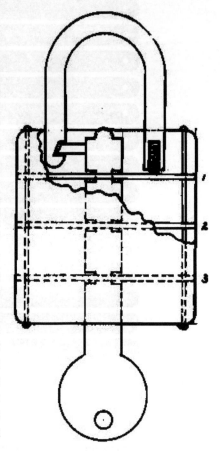

The mechanism is basically a flat hairpin type spring that latches into notches in the shackle. Only a portion, usually the tip of the key, actuates the spring latch. Turning the key spreads the spring latch apart, releasing the shackle. A pick for this simple design would only have to be a paper clip or wire with a small "L" bent on one end.

A pick key would also operate any such lock whose keyway would accept it. This is simply a key with all ward cuts opened up, leaving only the portion on the tip that is necessary to operate the latch spring. Manufacturers, in an effort to improve the security of this basic locking mechanism, have now added another spring latch with a ward between them. This design complicates efforts to pick it with a bent wire as previously done. However, a double headed pick key will do the job.

Pin tumbler locks are the most common type lock mechanism found today and are most commonly referred to the locksmith. To understand picking this lock you must be familiar, to a limited degree, with its mechanism. Manufacturers have dozens of versions of this mechanism, yet they are all basically mechanically the same. While they can have more or less, the average pin tumbler lock has five sets of tumbler pins consisting of a spring, top pin and bottom pin. The springs and top pins are usually the same length. The bottom pins vary in length to match the depth of the cuts in the key. When a key is inserted into the lock plug this set of pins is raised, compressing the spring. If the proper key has been inserted the bottom pins are all raised until they are flush with the diameter of the plug. This is also known as the shear line. At this point the plug is free to turn and release or activate whatever mechanism it is attached to.

To pick this mechanism we need to raise or manipulate these pins so as to allow the plug to turn. Most methods of picking this lock rely on the presence of tolerances. The tumbler pin holes seem to be the same diameter and also, in a straight line. They are supposed to be. However, if we were to measure each part of a lock with a precision measuring device, we would find that the diameters of the pins and holes may vary slightly from pin to pin and from hole to hole. Also the holes, instead of being in a perfectly straight line, will vary slightly from side to side. This variation may be only a fraction of a thousandth of an inch, but is enough to aid picking. Picking involves applying a very small turning force or torque to the plug and with a feeler pick, carefully probing each bottom pin to find the one or more that seem to be binding more than the rest. With the feeler pick slowly lift one of these until the top pin clears the shear line. At this time the plug may give slightly in the direction that torque is being applied. This operation is repeated on the remaining pins, at which time the plug will be free to turn.

Raking is another method of picking, perhaps the most often used because less skill is required as the lock opens more by chance than by skill. A rake tool has two or three up and down areas and is used in an in and out and up and down motion. The rake, together with the random motion, may at some unknown moment raise the bottom pins to the right level. If a small torque is being applied at this instant the plug will turn. Another form of raking involves using a diamond shaped tool. This tool is inserted all the way into the keyway and jerked out very fast. This motion tends to throw the pins apart because of inertia. This opens the area at the shear line permitting the plug to turn.

Rapping entails striking the body of the lock with a plastic, rawhide or other protective hammer in the opposite direction than the pins have to travel to reach the shear line. This technique has been used with some degree of success in opening padlocks where the latch dog was acted upon by this transfer of force rather than the pins themselves. Most quality padlocks have had design improvements to preclude opening by this technique. Any flat spring material ranging in thickness (.015 - .035 can he fashioned into a pick with the aid of a small grinder. One of the most common sources of such material is an automotive feeler gauge. These gauges have blades ranging in thickness from .001 to approximately .040 of an inch.

Technical Terms for the Beginner

BITTING

A bitting is the part of the key that actually engages the tumblers to activate the lock. They are often represented as a code which instructs how a key is to be cut by a locksmith. The bitting is usually a series of integers usually translated from a key code chart or from a bitting code list and the use of specially designed key machines.

Each digit in the bitting corresponds to a different cut or notch on the key and represents the depth at which the key must be cut. Each number in a bitting represents not only the depth of which the key blank is to be cut, but also the location of the cut on the key blank. Depending on the maker the bitting sequence can be from bow-to-tip (the bow being the larger handle portion of the key), or can be from tip-to-bow (as is in the case of Best Locking Systems and ASSA). A smaller number is typically a shallower cut on the key, but not always. ASSA bitting codes are reversed where the higher the digit, the shallower the cut. Locksmiths can cut to the code given when supplying a lost key or making a new restricted key copy.

BOLT STUMP

A bolt stump is a rectangular part of a lock located above the talon, and passes through the slot in the levers as the bolt moves. This part is welded or riveted onto the lathe in some earlier locks. Modern lock manufacture allows this to be machined in. Most are directly cast into the blank and then milled smooth for final use.

CHANGE KEY

A change key is a key on the lowest level of a master keying system. Change keys are also referred to as day keys. Typically change keys are issued to personnel that require access to one or two areas in a facility. By definition, a change key is not a master key. Change key can also refer to the specialised keys used to change the combination setting of a combination lock

KEY CODES

Blind codes - These are codes that require a chart or computer programme to translate the blind code to a bitting code, which is used to create the actual key. Most key codes are blind codes, and publication of code books or software are restricted to licenced locksmiths for security reasons. Some locksmiths also create their own blind coding systems for identifying key systems they have installed, or for customer identification and authorisation in high security systems. master keying charts also use blind codes for identifying individual change keys and masters within the system.

KEY BLANKS

A key blank (sometimes spelled keyblank) is a key that has not been cut to a specific bitting. The blank has a specific cross-sectional profile to match the keyway in a corresponding lock cylinder. Key blanks can be stamped with a manufacturer name, end-user logo or with a phrase, the most commonly seen being 'Do not duplicate'. Blanks are typically stocked by locksmiths for duplicating keys. The profile of the key bow or the large flat end often references an individual manufacturer.

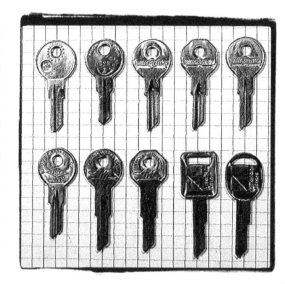

KEY RELEVANCE

In master locksmithing, key relevance is the measurable difference between an original key and a copy made of that key, either from a wax impression or directly from the original, and how similar the two keys are in size and shape It can also refer to the measurable difference between a key and the size required to fit and operate the keyway of its paired lock.

No two copies of keys are exactly the same, unless they were both made from key blanks that are struck from the same mould or cut from the same duplicating/milling machine with no changes to the bitting settings in between. Even under these circumstances there will be minute differences between the two key shapes, though their key relevance is extremely high. In all machining work, there are measurable amounts of difference between the design specification of an object and its actual manufactured size. In locksmithing the allowable tolerance is decided by the range of minute differences between a key's size and shape in comparison to the size and shape required to turn the tumblers within the lock. Key relevance is the measure of similarity between the key and the optimal size needed to fit the lock, or it is the similarity between a duplicate key and the original it is seeking to replicate. Key relevance cannot be deduced from a key code, since the key code merely refers to a central authoritative source for designed shapes and sizes of keys. Typical modern keys require a key relevance of approximately 0.03 millimetres (0.0012 in) to 0.07 millimetres (0.0028 in) (accuracy within 0.75% to 1.75%) in order to operate.

MAISON KEY SYSTEM

This is a keying system that permits a lock to be opened with a number of unique individual keys. Maison key systems are often found in apartment building common areas, such as main entrance or a laundry room, where individual residents can use their own apartment key to access these areas. Unlike a master key system, where each individual lock has one individual operating key and one common master key, Maison lock is designed to be operated by every key within the system. Because of the inherent lack of security in the Maison key system, some apartments prohibit their use. In such locations, access is usually facilitated by either a high-security key-controlled system or the use of electronic access control systems such as a card reader.

MASTER KEYING

A master key is intended to open a set of several locks. Usually, there is nothing special about the key itself, but rather the locks into which it will fit. These locks also have keys which are specific to each one (the change key) and cannot open any of the others in the set. Locks which have master keys have a second set of the mechanism used to open them which is identical to all of the others in the set of locks. For example, master keyed pin tumbler locks will have two shear points at each pin position, one for the change key and one for the master key. A far more secure (and more expensive) system has two cylinders in each lock, one for the change key and one for the master key.

Larger organisations, with more complex "grandmaster key" systems, may have several master key systems where the top level grandmaster key works in all of the locks in the system.

RE-KEY

Re-keying normally refers to the ability to change a lock so that a different key may operate it. Re-keying is done when a lock owner may be concerned that unauthorised persons have keys to the lock, so the lock may be altered by a locksmith so that only new keys will work. Re-keying may be done without replacement of the entire lock. Re-keying was first invented in 1836 by Solomon Andrews, a New Jersey locksmith. His lock had adjustable tumblers and keys, allowing the owner to rekey it at any time. Later in the 1850s, inventors Andrews and Newell patented removable tumblers which could be taken apart and scrambled. The keys had bits that were interchangeable, matching varying tumbler configurations. This arrangement later became the basis for combination locks.

SHEAR LINE

In a cylinder lock the shear line is where the inner cylinder ends and the outer cylinder begins. When a correct change key or master key is inserted in the cylinder, it will align the pin segments with the shear line and allow the cylinder to be turned. This break in the lock mechanism is a vital part of the lock picking process, as it allows a picked pin to hang while the others are being picked.

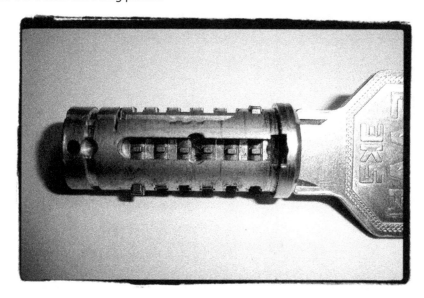

BASIC LOCK PRINCIPLES

Of all locks in general use today the simplest and least secure (from a standpoint of interior construction) is the warded lock.

Construction

A ward is merely an obstruction of some type which prohibits the key from operating unless a corresponding notch or ward cut is made in the key. There are a number of different types of wards and many ways they may be applied in the lock. The most noteworthy weakness in the construction of a warded lock is the fact that a pass key can usually be made out of one key that will fit any lock in the series, although there may be from four to twenty or even fifty so called "changes" in the set. Each little notch in the key bit is a case or cover ward. There is an obstruction sticking out into the keyhole on the lock case that refuses entry to any key except one having a longitudinal slot cut in it in the same location as the ward on the lock case. But if we file away the wedge shaped bit and make a key, we have a key blank that will fit into any of the various keyholes.

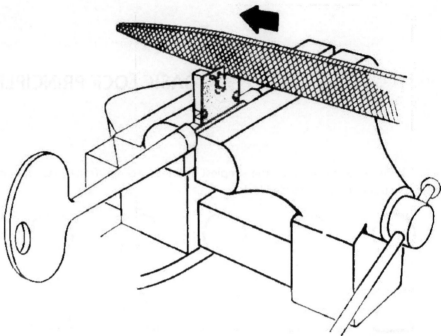

Naturally this would not be done if a key was being fitted to a single lock, for it would remove a large measure of the limited security of the lock. The notches in the keys are known as end wards. These obstructions are also attached to the case at the end of the key. These wards may be placed at the point it meets the ward to open this lock. If you cut the key down at all five different end ward locations and use a blank, we have a pass key. (10 case ward locations multiplied by 5 end ward locations gives a total of 50 possible

key changes) This is the weakness of the warded lock. The key needs only to be cut away past all the wards and it will open any similar lock. For this reason warded locks are seldom placed in service where any great amount of security is required. They serve quite satisfactorily to keep the family chest, sewing machine or piano. But beyond that their use almost ceases. Since a warded lock is in itself little protection against illegal or forcible entry, wards are most often used in conjunction with some other type of security, usually the lever tumbler plan. Perhaps the most commonly used straight warded lock today is the inexpensive padlock. This is usually designed with a flat or slightly corrugated steel key. Inside the lock are arranged a number of flat steel plates.

These plates, spaced at various intervals, provide a number of wards that restrict the key in its turning motion unless notches are filed on each side of the key everywhere there is a plate. Here again is the weakness of the warded lock. No matter how many ward plates are placed in the padlock, a key filed away would operate all of the locks in a given set.

Fitting Keys to Warded Locks

Because of their simplicity of construction, fitting keys to warded mortise locks is a relatively easy operation. In many cases it can be done without removing the lock from the door.

With the aid of a torch the shape of the keyhole can be examined, and a blank selected as near as possible to the exact size. An oversized, rather than an undersized blank, should be selected if the exact size required is not readily available. A large blank dressed down to the right size will fit snugly and never give trouble in riding too high or too low in the keyhole. Blanks that fit too loosely can cause trouble. The initial cut is the case ward cut if there is one. This should be carefully plotted on the key in its exact place before filing. The actual cut should be no larger than is necessary for easy entrance of the key.

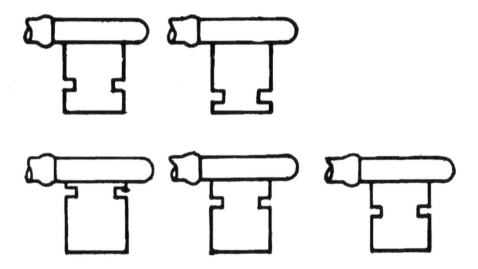

Once the key has entered the lock, the end wards can be located by smoking up the bit in a match or candle flame. When pressure is applied in a gentle attempt to turn the key in the lock, the wards will rub off the blacking on the key. Then when the key is filed where the bright spots appear in the blacking, the ward cuts have been made. Adjustments are then made if they are necessary to make operation faultless, and the key is ready for service.

For fitting keys to warded padlocks, the steps are even simpler, since there are no case wards. The simple side wards can all be located by the impression method. All that is necessary to start with is a blank of the proper shape. Because of their comparatively flimsy construction and low cost, it is usually not practical to repair warded locks. Cases will develop however, in which a locksmith will find it necessary to make repairs to a lock of this type for several reasons. While it is usually best to recommend a customer buys a new lock, it is not always possible to get one to fit the mortise or one that will satisfy the customer who thinks he really has a lock of great security. This is especially true in regard to imported chests, desks, strong boxes, etc., where replacements are not always available.

A careful inspection of the individual lock will show the cause of the problem. The main thing to look for is a broken spring. It may not appear broken when you inspect it, but look at the end away from where it is connected to the boltwork. Sometimes a spring will crystallise from constant use. Even a small piece breaking off may jam the lock if it falls into the moving parts. On certain lever type locks wards are used to aid in producing a larger number of possible key changes, as may be necessary in hotels etc. Often these wards break off or are removed by wear and must be replaced in order to keep other keys from operating the locks. Quite often wards can be replaced, or even new ones created, by boring a small hole and forcing a short length of a brass escutcheon pin into the lock case where the ward is desired. This method can be used only when the case is of cast or very heavy sheet metal. On thin metal cases wards are generally only indentations punched into the metal with some small sharp instrument.

Duplicating Bit Keys

Bit keys can vary in design from the very simple to the very complex. One design feature that occasionally appears on bit keys is a side groove. The purpose of such a groove is to restrict entrance into the lock by any key other than the one containing the proper groove. Whenever there is a side groove in a sample key you must file a corresponding groove in the duplicate. To locate the position of the groove, you should use a groove guide. This shows you where to file the side groove in the bit and it helps you measure the depth of the groove.

This how a groove guide is made:-

You must first form a soft metal strip into a mask. Bend the strip so that both flaps are approximately the same size.

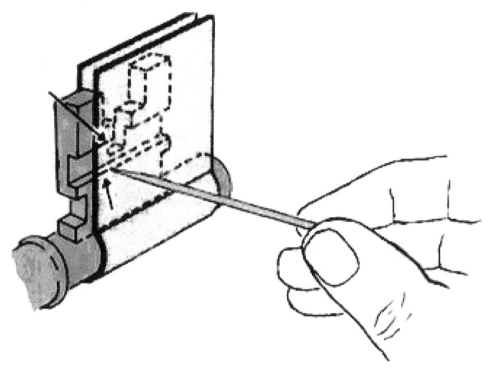

After you have pressed the strip snugly around the sample key, remove the assembly from your vice. Next slip the guide down along the post until you see the groove. Using a scriber make a scratch mark on the mask indicating the location of the top and bottom of the groove. Remove the groove guide from the sample key and slide it on to the post of your duplicate. Mark both edges of the bit. (If these marks do not stand out clearly, try smoking the bit of the blank first.)

Now you have marked the location of the groove, you must determine how deep it must be filed so that it will be equal in depth to the groove in the sample key. Stand one flap of your groove guide in¬side one of the walls of the groove in the sample key. Scribe a line on to the groove guide where it reaches the top of the groove.

This now becomes your tempo¬rary depth guide. Complete your key by cutting the groove into the blank using a ward¬ing file. Since the pur¬pose of the groove is to enable the key to pass an obstruction in the keyway, a groove which may be slightly wider or deeper than the one in the sample key is acceptable provided you do not weaken the key.

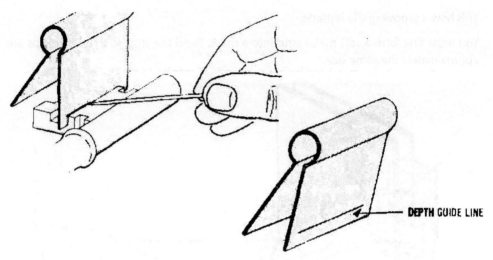

DEPTH GUIDE LINE

The temporary depth guide is usually all you need to determine the depth of the groove as the depth is not very critical on most bit keys. The groove just has to be deep and wide enough to pass into the keyhole. Some lock¬smiths prefer to use a slot gauge. This is nothing more than a piece of soft metal with a slot filed in it equal to the exact thickness of the bit at the bot¬tom of the groove. In the following illustration you see how the slot would be formed with a warding file. The slot is being widened just enough to let the groove of the sample key pass into it.

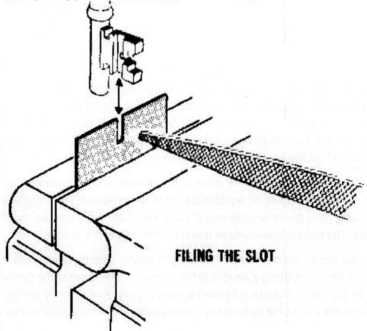

FILING THE SLOT

In the next illustration, you see that the groove in the duplicate key has been filed and the slotted gauge is being used to determine whether or not it is deep enough.

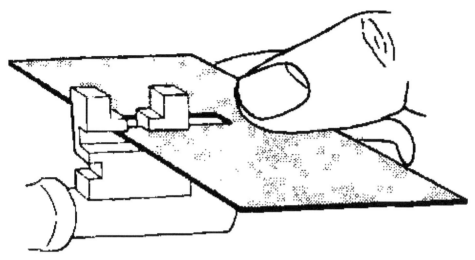

When comparing your sample and blank key for thickness of bit, ensure the blank you choose is not too thin. If you attempt to file a groove in a blank which is too thin, the strength of the bit may be severely weakened. The bit may possibly bend or break at the groove when you try the key in the lock.

Many locksmiths who duplicate a large number of bit keys find it easy to use a caliper or drill gauge to measure post size.

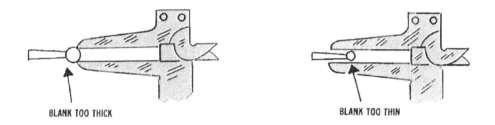

BLANK TOO THICK   BLANK TOO THIN

Using a Caliper

- Adjust the caliper jaws to fit gently over the post of the sample key.
- Remove the key. Be careful not to alter the setting.
- Attempt to slide the post of your blank key into the opening.

A proper diameter post will slide into the opening just barely dragging on the measuring surface as it moves. A post which is too large will not enter without forcing the opening wider. A post which is too thin will enter easily but wobble within the jaws of your caliper. (The allowable tolerance is 1/16" or .060")

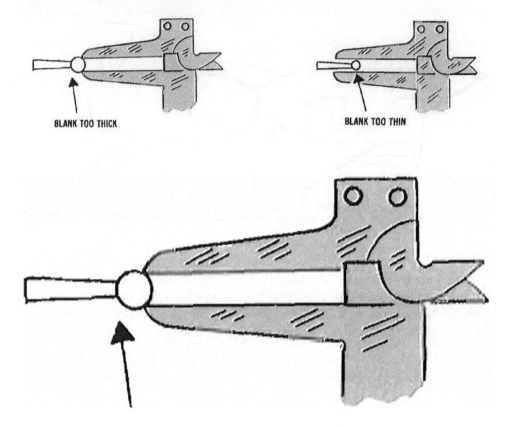

Using a Drill Gauge

- Attempt to fit the post of the sample key into the holes in the drill gage.
- Note the number of the hole in which the post fits best.
- Remove the sample key from the gage.

Insert Post in Proper Diameter Hole

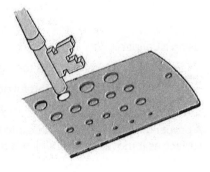

Try to insert the post of the blank key in the same hole. If the post fits as snugly as the post of the sample key use this blank to form your duplicate. (provided the other dimensions of height, width, length and thickness are acceptable) Do not use a blank whose post was too large to fit into the same hole as the sample.

If the post of the blank key enters the hole but fits loosely, find the hole in which it fits best. Note the difference in dimension between the two holes. If your blank is no more than 1/16" or .060" smaller than the sample, it may still be used.

A warded pick, also known as a skeleton key, is a device for opening warded locks. It is generally made to conform to a generalised key shape relatively simpler than the actual key used to open the lock. This simpler shape allows for internal manipulations. This style of pick can also be used to rip the lock and this is where the pick is placed at the back of the lock and then pulled out in one sharp fast ripping action.

The keys for warded locks only require the end section which is the one which actually open the locks. The other parts are there to distinguish between different variations of their locks. If you have a chest of drawers with a warded lock you can make a skeleton key for that type of warded lock by filing away all but the end of the key.

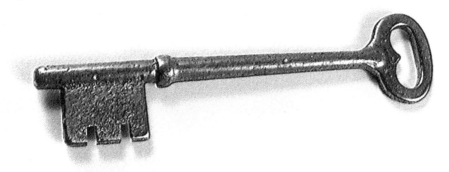

Warded locks are perhaps the most abundantly used locks throughout history. Being many centuries old, they date back well into the Roman and Egyptian eras. A warded lock is very simple in construction using wards or protrusions in the lock itself, to derive its security. Identification of a warded lock is simple. The familiar skeleton key is of typical style to that of a warded lock. The warded lock key hole is often stereotyped that you can look through the keyhole into the next room. This stereotype is often true.

The technique of using warding in the key hole has carried over to modern day locks, in which the keyhole warding keeps keys from different manufacturing companies from entering the wrong type of lock. Basically in a warded lock you have several cuts made in the key. These cuts are made to circumvent the wards of the lock, and ultimately open the bolt. If the cuts on the key do not match that of the internal warding the key is shunted, and therefore cannot turn.

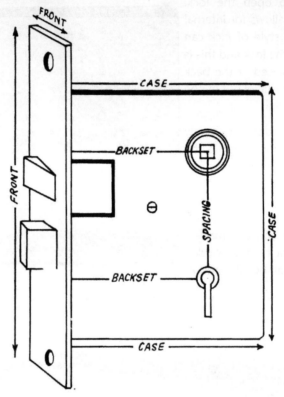

A warded lock has three active parts:-

1. the key
2. the internal lock warding
3. the bolt

Using these characteristics we can study how the lock works.

Key is inserted

- Does the key turn? If the key does not turn then the warding is hitting the key. You have used a wrong key.
- The key turns. This means that the cuts in the key were able to avoid hitting the

warding. This means you have used the correct key.

- Key contacts the bolt. The bolt opens.

Opening a Warded Lock

Warded locks are of such simple design and have such few moving parts that they are often called upon to operate in harsh environments such as near beaches or on ships. They may also be found on older style houses and on many padlocks. They are rarely found in newer buildings. Many times a Button Hook pick will open these locks. This pick can be easily made from a stiff coat hanger. Make several sizes, large and small, this way you are always prepared for any type of warded lock. This type of pick will circumvent much of the warding in the lock itself. When inspecting a warded lock you should be able to see the bolt that the key hits to disengage the lock. Make sure the top bend of the pick hits this bolt. Replace all warded locks that are entrusted with the security of important places and valuables. These locks offer very little security.

Warded Padlocks

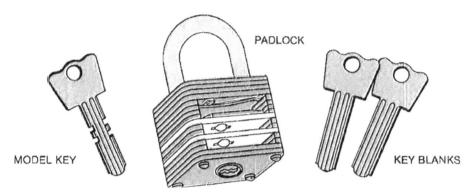

These are simple in design and inexpensive to produce. Warded padlocks are used in many homes, for lockers and bicycles. They can be various sizes. These locks are easily identified by the following characteristics:-

- The key hole is rather large and you can usually see the bottom of the key way when held under a light.
- The key is shaped much like a toy key. It has square notches cut on both sides of the key and the notches are symmetrical to each other (The notches match on both sides of the key).

These locks are easy to defeat. In order to defeat this lock you must understand how it works. Let us start with the key. The cuts in the warded type key are only there to either circumvent a ward, or hit the warding indicating that the wrong key has been used. The key must physically hit the bolt in order to disengage it. The wards will only stop the key if they are struck by a part of the key. If you remove the part of the key that has potential to strike the warding, you would have created a key that the lock thinks is correct. Master Lock and others use this technique to keep out other keys. To circumvent this, it is necessary to grind down the sides of your pass keys so they are relatively smooth, but not thin enough that it will break in the lock.

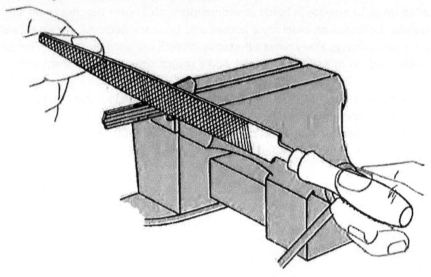

Impressioning: Making a key for a warded lock

Sometimes you do not want to just pick open a warded lock. Usually you will want a key if you need to open it again. Or perhaps you lost a key for a warded lock. Making a key for a warded lock is simple.

You need only four things:-

- warded lock key blanks
- files
- matches, candle or lighter
- patience

Again we refer to the warded lock operating principles. The wards hit the key to stop it correct? Using this fact you follow this procedure:-

- Hold key over the candle or match until a thick layer of carbon builds on all the sides of the blade.
- Insert key into the keyway and turn in the direction to throw the bolt.
- Remove the key and look at the places where the carbon has been removed.
- Gradually start filing these spots only taking off a little at a time.

- Repeat steps 1-4 until you notice that the carbon is no longer being displaced.

It is essential that you start taking off less and less with the file as you go deeper into the key blank. Be sure that you file all the places that have had the carbon removed. Eventually you will end up with a key that will work the lock.

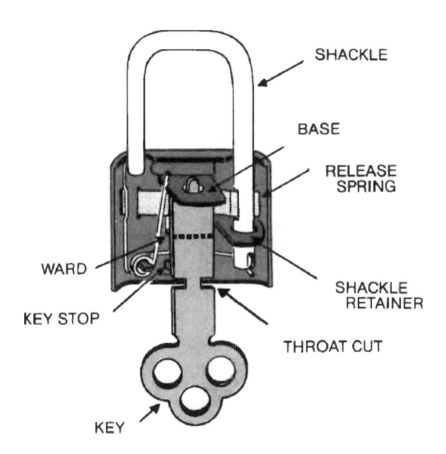

PIN TUMBLER LOCKS AND BYPASSES

Locks embodying the pin tumbler principle are not only the most widely used and universally accepted throughout the world, but are also generally recognised by expert and layman alike as the highest type security available.

The greatest advantage of the pin tumbler mechanism, apart from its burglar resistance, is its adaptability to any type of locking problem. Pin tumbler locks are available for almost any imaginable use. You find them everywhere on dog collars, money chests, public buildings, residences, automobiles and even bicycles.

The pin tumbler lock, unlike any of the other types, depends for its security on a number of round pins or tumblers operating in a cylinder not necessarily a permanent part of the lock itself. The ordinary mortise door lock of this type, for instance, is used with a cylinder. The cylinder, although an integral part of the lock, is an individual unit and is readily removed for repair or service without disturbing the lock itself. This feature is a particular advantage in large buildings and other institutions when tenants move, authority is changed or access restricted.

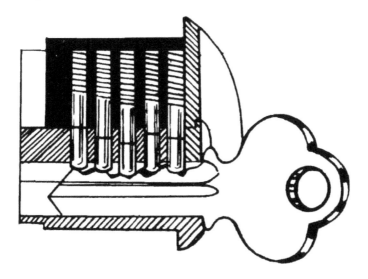

Operation of the pin tumblers should be quite clear after studying the illustration. Each tumbler or pin is divided, in the ordinary cylinder, into two parts. The upper part, flat on both ends, is known as the driver. The bottom part, or pin, is either rounded or slightly pointed on the lower end to fit the V shaped grooves or cuts in the key. A separate coil spring above each tumbler constantly forces it downward. When the proper key is inserted the various depths of the cuts in the key compensate for the different length of the pins, and the dividing point between each two pin segments is brought into line with the top of the core. This allows it to rotate in the shell. When the core turns it carries with it the cam.

Locks of the pin tumbler type do not depend for security on the tumblers alone. Each manufacturer has a number of different keyways, or keyhole, shapes. These, combined with an almost countless number of tumbler variations, provide a great reservoir of key changes. Most standard sized pin tumbler cylinders have five tumblers, although as few as three and as many as eight are in general use. Assuming that there are ten different lengths each pin can be (most locks have at least that many), and five tumblers, then there would be 100,000 (105 or ten to the 5th power) different theoretical key changes. But twenty or so leading manufacturers each have from five to a dozen or twenty different keyways, so our 100,000 becomes 2,000,000 and then 20,000,000 if we agree that each maker uses an average of ten keyways.

The possibilities for security then are great. There is roughly one chance in 20,000,000 of anyone else having a key to fit a given pin tumbler lock. Lock manufacturers have shaped the keyways of their lock cylinders as scientifically as possible. Milled or corrugated keyways offer maximum resistance to unauthorised entry. Generally speaking, the smaller the keyhole, and the more obstructions in the keyhole there are, the harder it is for even an expert to open the lock. The keyholes should not be made too small or the soft metal keys would be continually breaking off in the lock.

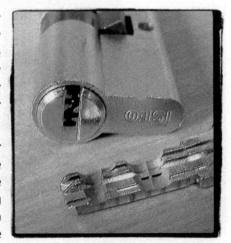

Nationwide standardisation of cylinder dimensions and thread sizes has greatly aided the locksmith. In many cases, where it is necessary to remove a cylinder for key changing or repair, a substitute or loan cylinder can be easily fitted to the lock while the regular cylinder is in the shop.

Standard mortise cylinders or cylinders for locks mortised into the door, whether there is a single cylinder or one on each side of the door, are removed from the lock by loosening a cylinder set screw in the edge of the door. In some locks this screw is concealed by a false front or armoured front, which must be removed before the set screw is available. After this screw is loosened the cylinder is easily unscrewed from the lock case. Night latch or rim lock cylinders are held in place by connecting screws through a back plate. The lock case itself usually has to be removed before these are accessible. Cylinders on drawer and cupboard locks, automobile latches, padlocks and many other types are not removable in this way and are generally contained in the lock itself. In such cases the cylinder is available for key changing or repair only after the lock is almost completely disassembled.

Fitting Keys to Pin Tumbler Cylinders

Naturally the most common reason for opening a pin tumbler cylinder is to fit keys. The customer may request keys fitted to the lock as it is, or he may wish the combination changed first and new keys fitted. In either case the plug must first be removed from the cylinder. First take off the cam or tail piece, whichever the case may be. Then, if the old key is available, the plug can be slid out of the lock being closely followed by a follower or following tool to keep the springs and drivers in the shell from spilling.

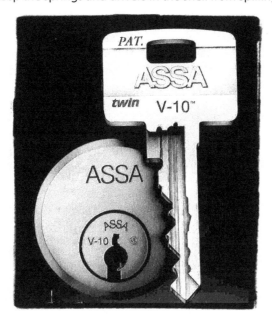

Setting up a Cylinder

When it is necessary for any reason to remove the upper pins or drivers from a pin tumbler cylinder much time can be saved in replacing the springs and drivers by following the directions below.

First insert the following tool in the empty shell. Pull it about halfway out until three pin holes are visible by looking into the front of the plug hole. With tweezers place a spring in the third hole, then a driver on top of it, pushing the driver down until the following tool can be slid over it to hold it in place. Proceed in the same manner for the second and first hole. Then pull the rod through the cylinder to the opposite end and set up the other two or three pins in the same way from the back of the cylinder, starting with the fourth pin from the front or bow of the key.

Occasionally a cylinder will be brought to the locksmith with the explanation that there are no keys to fit it. The first problem in this case is to get the cylinder open so the plug can be re-removed. Since there is no key, a professional method must be applied.

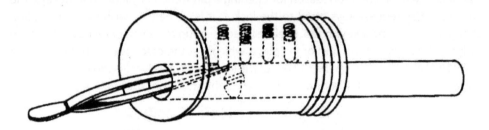

One way to crack a cylinder under such circumstances is to insert a short length of very fine flat steel spring wire (ask your watchmaker for a piece of broken watch spring) between the plug and the shell from the back of the cylinder. Meanwhile work a blank key gently back and forth in the keyhole to move the pins up and down. As each division point is brought even with the spring wire the steel can be pushed in a little farther, until all the pins are separated. Then the plug can be turned very slightly, the spring withdrawn, the following tool set in place and the plug removed as if it had been opened with a key.

Sometimes, especially on newer cylinders, there is not enough space between plug and shell to permit the flat steel to enter. Then it is necessary to rap the cylinder to open it. In using this method the cam is removed and the cylinder held in the left hand with the keyhole down, and the thumb exerting an outward pressure on the plug. A few sharp blows of a light wood or plastic hammer will sometimes jar the pins into the open position. Do not let the plug slide out too far before it is turned and above all do not hammer the cylinder with a metal hammer.

This method is by no means foolproof and depends for its success largely on the skill of the operator in acquiring the knack of wielding the hammer, and on the setup of the pins in the lock. Sometimes even a beginner can crack the cylinder with the first or second blow. Other times an expert can hammer till his hand is black and blue without any luck. When the cylinder has been opened by one of these two methods the next step is relatively easy. If the combination is not to be disturbed a blank is selected to fit the keyhole. Round-bottomed grooves are filed in it under every pin tumbler segment until each is exactly level with the top of the plug. When this has been done, the hills and valleys in the key should be smoothed out slightly so as not to leave any sharp points on the key that will hinder it entering or leaving the cylinder. No slope on a pin tumbler key should be at an angle greater than 45 degrees from the horizontal.

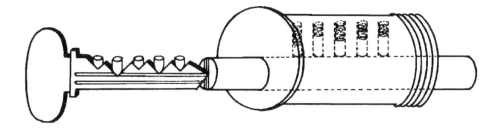

Repairing Pin Tumbler Locks

Probably 90% of the trouble locksmiths are called upon to repair in pin tumbler locks is confined exclusively to the cylinder itself. But the simplest source of trouble, although a common one, is a loose cylinder set screw. This lets the whole cylinder turn when the key is inserted, instead of holding the cylinder in an upright position while the plug rotates inside it. Quite often a key is hard to insert in the lock or it may be difficult to remove. A number of factors can contribute to this situation. First check the key for sharp points or rough spots. Then remove the cylinder and see that the cam screws are tight. If they are loose, or the plug is too long because of wear on the cylinder, the trouble has been located. Remove the cam and very slightly shorten the plug by a gentle stroke or two of a flat file. This will make the plug fit tightly and prevent its backing out when the key is withdrawn.If the operation is still not satisfactory, soak the cylinder in petrol or cleaning solvent. It may be that someone has gummed it up by oiling it.

After the petrol has thoroughly dried out, a shot of graphite should do the trick.This petrol cleaning is especially desirable at regular intervals on public buildings, hotels, and apartments where brass polish accumulates in the lock. Spare tyres and rear door locks in cars are susceptible to failure at critical times because of road dust and oil film. It can best be accomplished with an ordinary blow torch. Pump up the pressure and direct the small steady stream of liquid petrol into the keyhole. Run the key back and forth a few times to cut the dirt and then rinse with another stream of petrol.

Poorly fitted or worn out keys create their share of trouble in these locks just as in any other kind. When new keys are fitted to an old cylinder it is always a good plan not only to replace the lower pins, but the upper ones or drivers, and even the tumbler springs, unless they retain sufficient tension to actuate the pins properly. A dead pin tumbler spring, especially on a lock not mounted on an upright door, can cause all kinds of trouble.

Pin tumbler locks, exclusive of the cylinders, are usually a great deal more complex than any other type. Perhaps residential front doors give the greatest amount of trouble. Those equipped with handles on the outside are especially vulnerable to operating irregularities. After a few years the thumb piece bends from use and then the latch does not retract far enough to open the door when the thumb piece is depressed. Usually straightening or reinforcing the thumb piece puts the lock in operation.

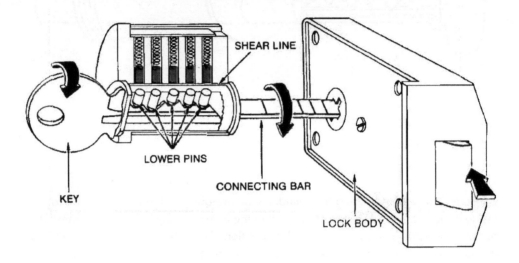

Cylinder locks equipped with knobs on both sides frequently have spindle trouble. The swivel on the jointed spindle has to be kept exactly in the centre of the two piece lock hub or the night latch feature of the lock will not work properly. Wear to knobs or loose set screws often move the spindle either in or out. This is easily corrected by the locksmith. Merely centre the spindle, being sure to unscrew the two sections of it far enough to insure their not jamming together when the knob is turned.

Caution when Working on Pin Tumbler Locks

Pin tumbler lock cylinders should always be lubricated with powdered or flaked graphite, never with oil. Dust and grit will stick to any liquid in the lock and soon cause gumming. Then tumblers move sluggishly, keys do not fit properly and may not operate the lock at all. Graphite should not be used too freely. An overdose can be harmful, just as a medicine could be to a person. It is messy to work with, so use it sparingly and wisely. After the graphite is applied work it in thoroughly by running the key quickly back and forth in the lock. Probably the best way to use graphite is in an air gun type of applicator. These are available at most hardware shops.

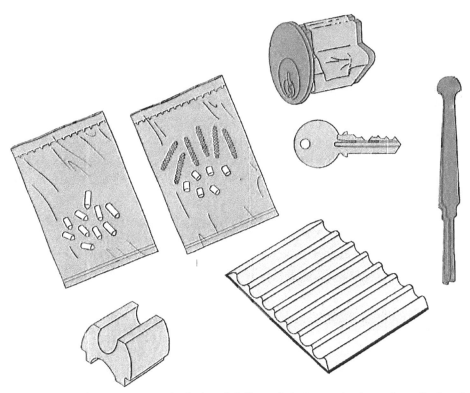

Some manufacturers are in the habit of drilling all their standard mortise cylinders for six pins, even though stock cylinders are ordinarily equipped with only five. Pay particular attention to this possibility in disassembling a cylinder. If there is one tumbler missing when you take out the plug, by no means put in one when you change the combination unless you also add a spring and driver in the shell to actuate the extra tumbler.

Sometimes when a cylinder is dismantled it will be discovered that there are more than two pin segments in each tumbler. Such a situation indicates that the lock is master keyed. Care should be taken not to disturb the arrangement of the tumblers unless required.

When a cylinder is cracked, either by rapping or springing with a fine wire, it is necessary to be cautious in removing the core. If the core is slid straight out the front four tumblers in the shell will spring into the four rear holes in the plug, and the cylinder will be locked part way open. Whenever a core is removed it should be turned slightly before being withdrawn to insure against this difficulty.

Occasionally after a pin tumbler combination has been changed it will be found that it is difficult to insert or remove the key. This condition is usually caused by too long a pin segment for a driver. If for instance the cut in the old key in No. 4 tumbler position was very shallow, there would have been a relatively long driver in the No. 4 hole to compensate for it.

But if in the new combination the No. 4 tumbler cut is very deep, the long pin and long driver would together take up so much more room that there would be no room for the tumbler spring. Replacing the driver in this case with a considerably shorter pin segment should alleviate the trouble, but the tumbler spring should also be inspected for damage from crushing.

Care must be used in working on pin tumbler locks to use the correct size tumblers. Too small a segment may cause wear and jam in a crosswise position, defying all efforts to open the lock.

Make sure you understand a customer's instructions before he leaves the shop. If he does not want a lock changed, and you change it by mistake or misunderstanding, you may cause him a great deal of trouble. Always give the customer back all the parts for his lock. Do not leave the cylinder rings or escutcheon plates on the bench.

In switching cylinders in mortise locks to save time and trouble in combination changes, avoid putting standard Yale, Corbin, Sargent, or other .51 inch plug cylinders in Russwin or other locks intended for cylinders having a larger core. The centre of revolution or axis is different in these locks, and a standard sized cylinder will not satisfactorily operate the lock intended for a different cylinder. The new Russwin cylinders have a special shaped cloverleaf cam which cannot be used in other standard locks.

[[IMAGE INSTALLING PINS IN PIN TUMBLER LOCK]]

Keys for pin tumbler cylinders should be, whenever possible, cut from key blanks from the manufacturer of the lock itself. Such a blank, known to the trade as a genuine, eliminates many of the dangers of imitation blanks. Some imitations are good, others are not.

Removing cylinders from mortise locks is sometimes a problem, especially if the cylinders refuse to come out after the set screw is loosened. Never put a pipe wrench or pair of pliers on a cylinder. Such procedure not only mars the cylinder and scratches the door, but brands the wrench wielder as a rank amateur if not a butcher.

If a cylinder will not unscrew, any of the following operations should aid in its removal. Place a scrap key or a key blank in the keyhole and, if necessary, use the pliers on the key. If it still turns hard loosen the escutcheon plate, if there is one, and remove the screws that hold the lock in the door. A gentle tapping on the face of the lock case with a light wood or plastic hammer should bring the cylinder out without any further trouble. Always put a thin coating of cup grease on the first few threads of a cylinder when replacing it to facilitate easy removal in the future.

in preparing keys for changing pin tumbler lock combinations, there are two things to avoid. Never let the first cut near the bow be more than of medium depth. To cut away the key at this point to a No. 8 or No. 9 cut materially weakens the key. Many people have a tendency to exert a turning pressure on the key before it is all the way in the lock cylinder. A key with a low cut near the bow would easily twist off, leaving a piece of key in the lock.

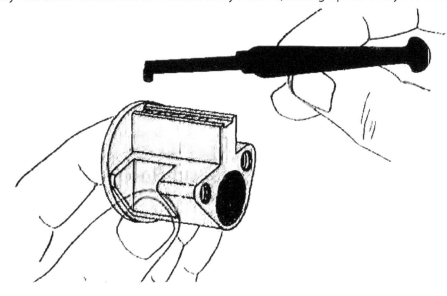

DISK TUMBLER LOCKS AND BYPASSES

One of the many accomplishments of the lock industry in recent years has been the de¬velopment of the inexpensive disc tumbler lock.

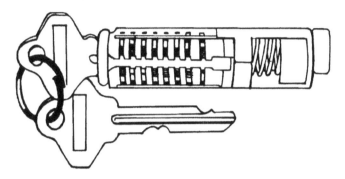

Ranking somewhere below the pin tumbler and better grade lever lock in security, the disc tumbler principle has been applied to a wide variety of locking problems. Many cars, slot machines and vending machines have been equipped with disc tumbler locks. Popular because of their low cost, these locks have found their way even to the inexpensive grades of builders' hardware for residential construction.

The Disc Tumbler Mechanism

The disc tumbler principle, while essentially similar to the pin tumbler idea in operation is very different in construction features. Disc tumblers are shaped and stamped or cut out of thin sheet metal. They are placed in slots in a core with the hooks alternating from side to side and a small coil spring under the hook of each tumbler to exert a constant upward force. The key looks almost exactly like one for a pin tumbler lock, except for its smaller size. When the key is inserted it goes through the punched-

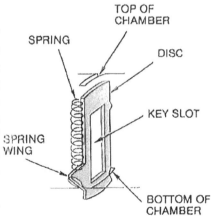

57

out hole in each tumbler, setting the discs flush at each side of the core and permitting it to ro¬tate in the shell.

Construction and Security of the Disc Tumbler Lock

Most disc tumbler lock cylinders are made as die castings or cast from an alloy inferior to brass. For this reason they are hard to repair. The usual five tumblers do not give as many key changes as a pin tumbler lock with the same number of tumblers, because there are only four or five variations in each tumbler. Security is limited to a theoretical 1024 possible changes, but in actual manufacturing practice this is fur¬ther reduced to about 500 or even 250. Disre¬garding the loss of security from a looser fitting plug the disc tumbler lock has only about 1/200 or even 1/400 the security of a pin tumbler lock of similar size and shape.

Instead of being fastened in the cylinder shell by a cam or tailpiece, the disc tumbler plug is usually secured by a spring clip or retainer spring near the back of the plug. When the key is inserted a small wire can be introduced through a hole in the shell to depress this retainer and the core can be removed.

Repairing Disc Tumbler Locks

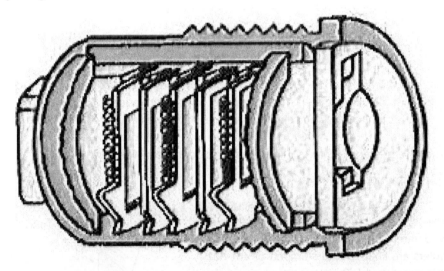

Repairs to disc tumbler locks consist almost entirely of replacing defective springs or tumb¬lers. Manufacturers have made new replace¬ment cylinders for these locks available at such low cost that no locksmith, even a beginner, can afford to spend shop time attempting repairs. In most cases it is more costly for a customer to have two keys fitted to a disc tumbler padlock than it would cost to replace the lock, including new keys. Some manufacturers at one time adopted what is known as a crushable disc tumbler for repairing locks of this type. These crushable tumblers were especially popular when most cars had disc tumbler door, ignition and tyre locks. The tumbler was really an adjustable one, which was the same size for all cuts in the key. After the key was inserted the tumblers were crushed into shape. While the idea of carrying only one simple size of tumbler was very convenient, the crushable tumbler was unpopular because it was impractical. In most cases the tumblers crushed unevenly causing binding or sticking, and in nearly all instances crushable tumblers left sharp corners sticking out to wear the soft metal of the lock shell. When ground away this formed a metallic dust which interfered with proper action of the lock. At its best the disc tumbler lock is a poor substitute for a pin tumbler mechanism.

The Disc Tumbler Lock today

Around 1930 disc tumbler locks were all the rage for cars, vending machines and padlocks. There has been a gradual trend away from the disc tumbler mechanism. Experience proved they were not strong enough to resist jimmying or secure enough to prevent unauthorised entry with keys that were not intended to fit them. One company today manufactures a genuine pin tumbler padlock with a laminated steel case that sells for less than a competitive die cast disc tumbler padlock. Common sense dictates that the pin tumbler lock will continue to gain favour under such circumstances.

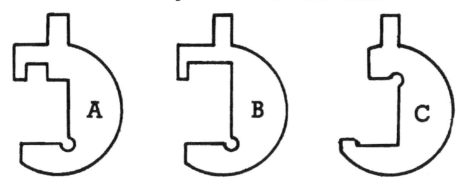

The Schlage Wafer tumbler Lock

The greatest improvement to the disc tumb¬ler principle has been the intro¬duction to the market of a wafer tumbler cylinder lock by the Schlage Lock Company of San Francisco.

In this lock the die cast cylinder has been discarded in favour of a tubular steel plug revolv¬ing in a steel shell which is built inside the knob of the Schlage unit-type lock. The wafers or tumblers of this lock are made of considerably heavier metal than the discs in previously men¬tioned types. As an added security feature, the number of tumblers has been increased to eight, with provision being made for almost one thousand absolutely different key changes. The interior construction of this "A" type Schlage lock is such that the key to one can seldom be made to oper¬ate another.These relatively inexpensive locks are proving very popular on modest resid¬ences, motor courts, tourist cabins, beach homes, etc., a few simple service instructions cannot be out of place.

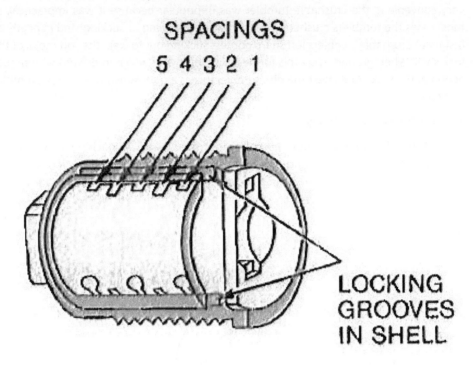

SPACINGS

5 4 3 2 1

LOCKING GROOVES IN SHELL

The core of this lock uses three different types of wafer tumblers. Type A, having a restricted opening, is used only on the eighth or last tumbler position on the tip of the key, which is cut either on the right or left in all cases, never left blank.Wafers B and C are used in the other seven tumbler positions on ordinary stock locks, and on special jobs where a master key controls the whole set. Note that besides the end, or eighth cut, no Schlage key of this type (except a master key) is ever cut on more than three of the remaining seven tumbler positions. These cuts can all be on one side of the key, or they can be alternated in any conceivable man¬ner, but there should never be more than a total of four cuts if security is to be maintained.

When you have mastered the art of picking open pin tumbler cylinders, you will have no difficulty in picking disc tumbler locks. One type of pick is diamond shaped. This pick is used quite successfully on disc tumbler locks. However, it does its job in a different manner from the other two picks that you have been using.

When the pick is inserted into the keyway, the tumbler is raised or lowered. With this pick, it is possible to feel when the tumbler reaches the shear line merely by sliding it along. This type of pick can also be used as a rake if necessary.Remember, lock picking is not always the easiest way to do the job. In the case of pin tumbler locks, the only advantage is to open the lock in an emergency where keys have been lost, and the door is locked. Some types of locks are very difficult to pick. If you are in the lock¬smithing business merely to satisfy your ego, you can undoubtedly succeed in pick¬ing them. But common sense and business principles should guide you. There is no point in wasting hours in picking a lock if the job can be done in a matter of minutes by some other method.

More on opening the wafer or disk tumbler lock

There are two kinds of keyway and keys to fit each in this system. The operation is simple. The rearmost wafer is called a master wafer, and is usually (in rest position with no key inserted) protruded from the surface of the plug by spring pressure, just like conventional disc tumblers. The key must have no bitting at that point where it contacts the wafer if the key is to wedge the wafer and retract it into the plug. The next seven parallel slots are filled with either a series (also no bitting on the key in order to retract them) or combination (requires a cut in the key at that point or else the key will wedge them out upon insertion and they will protrude from the plug surface) of wafers. By mixing the series and combination wafers in the slots, a variety of key combinations are possible. Remember though, the wafers only have one depth of protrusion, and the key only has one depth of bitting - all or none. In addition, the extreme tip of the key acts on the one master wafer and retracts it upon insertion.

The other type of the key tip is cut away to allow complete key insertion. If the key is not cut away opposite the master wafer side, the key will not bottom in the lock, and the bitting and wafers will misalign. To ascertain which type of key you have, place it with the central v groove, (the indented part) toward you, the tip to the left, bow to the right. If the key tip is cut away above the v groove, it is type one key, if the tip below the "v" groove is cut away, it is type two key. Observing which type will be of value in bypassing the lock.Identifying a wafer tumbler lock is very easy, due to the distinctive keyway and the fact that most applications are SCHLAGE key in knob exterior lock sets. There are two methods of bypass, one of them requiring specialist tools. The easy way is to insert a tension wrench to fit, use a lifter to count back to the eighth wafer, and retract it toward the body of the plug. A fair amount of tension may be necessary to keep it retracted. The next move is to sort through all the wafers in turn with the lifter, finding the series wafers and manipulating them, and leaving the combination wafers alone since they are already retracted when at rest. The combination wafers will show slightly more resistance when moved. The second method involves the use of a set of keys that has been cut down so that the tip retracts the master wafer, but the rest of the key is skeletonised so as not to operate any combination wafers. When this tool is inserted and tension applied, then a separate straight tool is inserted in the portion of the keyway left open by the key, and the series tumblers are retracted individually. This tool set is available from suppliers.

One other method of bypass involves inserting the proper type key blank into the lock, twisting hard, then maintaining tension and withdrawing the key almost all of the way, then slacking off the tension bit by bit until the combination wafers slip back into place. At that point, the plug should turn (theoretically), but this does not always work. Raking and other forms of vibration bypass do not apply here.One other form of bypass this lock is particularly vulnerable to is taking a quick impression of the key (wax or clay) or even a quick glance at the key, and cutting a duplicate, since key cut depth is not critical. The feel of bypass and the technique involved are really trivial, so do not spend too much time on these locks but know them thoroughly. Wafer locks are used on filing cabinets, lockers, cars, garage doors, desks and wherever medium security is required. The only wafer tumbler lock in common use that is difficult to pick is the side-bar wafer lock. It is the most popular type of car lock. This lock is of different design than most other locks and offers much more security than a regular wafer tumbler lock, or even a pin tumbler lock. The side bar lock was used mostly on General Motors cars and trucks since 1935. It is used on ignitions, doors and boot locks. Side bar locks are hard to pick because you cannot feel or hear the tumblers align with the cylinders breaking point. A spring-loaded bar falls into place to allow the cylinder to turn when all of the tumblers are aligned. There is no way to tell when that happens.   One learns to sense the bar while picking so that it seems to fall into place by itself.

But for beginners look down the keyway and locate the side groove of any of the tumblers using a pick as a searching tool. Drill a small hole in the shell of the lock above the bar which is above the grooves on the tumblers. Using an L-shaped steel wire, put pressure on the sidebar and rake the tumblers using a tension wrench for cylinder rotation and the lock will open. If you are going to be successful at opening side bars you will do it within two minutes. Otherwise you will cause unnecessary wear on your picks, not to mention wasting your time. Ford car locks can be relatively simple to pick. They have pin tumblers and remember that the door locks turn counter clockwise. Most other car locks turn clockwise. If you are not sure, remember this. If the tumblers will not catch at their breaking points you are going in the wrong direction with the tension wrench.

Wafer locks are a cinch to pick if you have learned how to pick pin tumblers. Just remember that wafers are thinner than pins and there is less distance between them. Generally you need less tension-wrench pressure with these locks, yet car locks can be quite stubborn and require a great deal of tension. Any heavily spring-loaded cylinder needs a substantial amount of tension. As a rule wafer locks need less play with the tension wrench than with pin tumbler locks. But if you find yourself having difficulty in opening these, you may try a little tension-wrench play. Usually they will not pop open like pin tumbler locks and they just slide open. You do not get the warning that a pin tumbler gives before it opens because there is less contact area on the wafer's edge than on a pin, so the sense of climax is reduced with these types of locks. Still, they open quite easily.

Another technique that can be used is scrubbing, a technique that can quickly open most locks. The slow step in basic picking is locating the pin which is binding the most. Assume that all the pins could be characterised by the same force diagram. That they all bind at once and all encounter the same friction. Now consider the effect of running the pick over all the pins with a pressure that is great enough to overcome the spring and friction forces, but not great enough to overcome the collision force of the key pin hitting the hull. Any pressure that is above the flat portion of the force graph and below the top of the peak will work. As the pick passes over a pin, the pin will rise until it hits the hull, but it will not enter the hull.

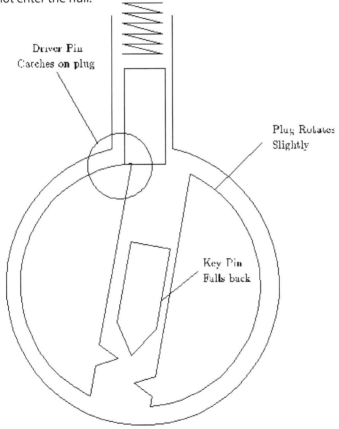

The collision force at the sheer line resists the pressure of the pick, so the pick rides over the pin without pressing it into the hull. If the proper torque is being applied, the plug will rotate slightly. As the pick leaves the pin, the key pin will fall back to its initial position, but the driver pin will catch on the edge of the plug and stay above the sheer line. In theory one stroke of the pick over the pins will cause the lock to open. In practice, at most one or two pins will set during a single stroke of the pick, so several strokes are necessary. Basically, you use the pick to scrub back and forth over the pins while you adjust the amount of torque on the plug. You will find that the pins of a lock tend to set in a particular order. Many factors affect this order, but the primary cause is a misalignment between the centre axis of the plug and the axis on which the holes were drilled. If the axis of the pin holes is skewed from the centre line of the plug, then the pins will set from back to front if the plug is turned one way, and from front to back if the plug is turned the other way. Many locks have this defect. Scrubbing is fast because you do not need to pay attention to individual pins. You only need to find the correct torque and pressure. The exercises will teach you how to recognise when a pin is set and how to apply the correct forces.

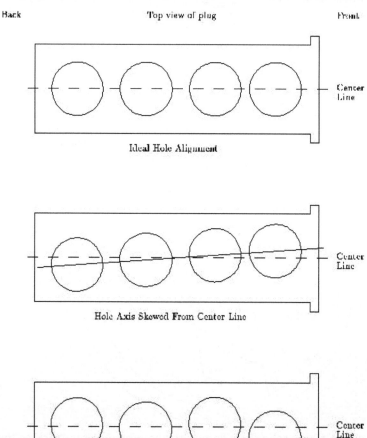

64

Sequence of Scrubbing to Open a Lock

- Insert the pick and torque wrench. Without applying any torque pull the pick out to get a feel for the stiffness of the lock's springs.

- Apply a light torque. Insert the pick without touching the pins. As you pull the pick out, apply pressure to the pins. The pressure should be slightly larger than the minimum necessary to overcome the spring force.

- Gradually increase the torque with each stroke of the pick until pins begin to set.

- Keeping the torque fixed, scrub back and forth over the pins that have not set. If additional pins do not set, release the torque and start over with the torque found in the last step.

- Once the majority of the pins have been set, increase the torque and scrub the pins with a slightly larger pressure. This will set any pins which have set low due to bevelled edges, etc.

LEVER TUMBLER LOCKS AND BYPASSES

The lever tumbler mechanism was undoubt¬edly the first satisfactory arrangement for pro¬viding security in early locks. In modem manufacturing practice lever tumbler action is used in a great variety of locks for hundreds of different purposes. From the relatively small two-lever drawer lock to the sixteen-lever double custody safe deposit box lock. There is a lever lock for almost every need.

The lever tumbler principle, when applied to a lock of good quality, carries a degree of security meeting or surpassing any ordinary re¬quirements for general locking purposes. Locks of this type can in most cases can be master keyed, keyed alike or all keyed differently.

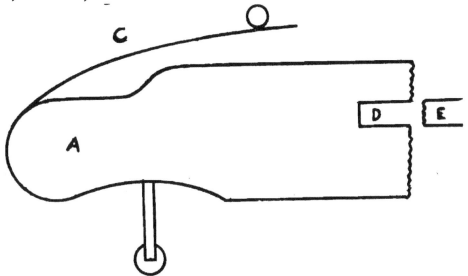

The lever tumbler mechanism itself is rela¬tively simple compared to some of the more com¬plex arrangements.

Each flat plate, or lever, hinges at a fixed point and is held down against a stop by pressure of a flat spring. Each lever has a gate filed in it, but these gates are all located at different places. When the proper key is insert¬ed and turned notches of various depths raise all the levers whatever distance is required to line all the gates up exactly opposite the post on the bolt. Then, when the key is turned a little farther, a corner of the key catches the bolt and slides it back. Since there is no resistance to the post entering the gate, the lock is opened. But if the key is not the right one for the lock, and even one gate does not line up to let the post slide into it, the lock cannot be opened.

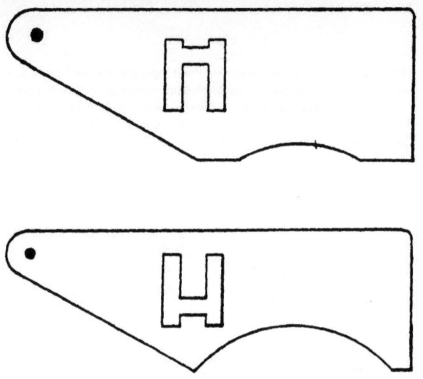

Whether locked or un¬locked the post is always snug in one of the two positions. When the key is out of the lock, the bolt cannot move. This feature is nec¬essary on all locks where it is desired to remove the key in the unlocked position. Otherwise vibration or slamming a door might jar the bolt half-way out, and in that position the key would neither lock nor unlock it.

In the case of the safety deposit box lock, and of certain mail box locks, it is intended that the user shall not remove his key until he has safely locked up the box again. As a safety feature manufacturers have made it im¬possible to leave these boxes unlocked. The key cannot be removed until the bolt slides out to the locked position.

Until fairly recently levers were sometimes so arranged that they pivoted on two different points instead of all on the same pin. When this is the case the key fits in between the levers and separates them to the proper degree to open the locks. There was a slight advantage to the user in the double bitted type. The key did not have to be inserted in the keyhole in a certain position. Either right side up or wrong side up, it still oper¬ated, and in theory each side of the key had only half the wear.

But the disadvantages were many. Keys were very difficult to fit and hard enough to duplicate. Levers were easily mixed up when the lock was disassembled. Present day standardisation of manufactur¬ing practices has all but eliminated this type of lock, except for interior doors on safes where bulk is an asset rather than a hindrance.

Key fitting to lever locks is in many cases greatly simplified by a small window in the lock case just over where the post on the bolt meets the gates on the levers. It is only necessary to remove the lock from the door, select or prepare a suitable blank, and notch it to fit by watching the action of the blank through the little window.

Since lever locks contain a great many loose parts and are often hard to re-assemble, they should ordinarily not be opened for key fitting, but only for repairs. By smoking up the blank the lever locations can be easily spotted. Get the lock in the locked position, with the bolt extended. Start with the lever closest to the window and carefully, a little at a time, file down the blank at lever position number one un-til the bolt post is exactly opposite the gate. Then follow the same routine for the second lever, and the next, until the gates on all the levers form a straight line. Try turning the key gently. If it binds, look for the cause. You will probably find that each lever has to be a fraction deeper. If it turns over but seems to stick, re-smoke the key and unlock it again. The place it binds hardest will show as the shiniest spot. A gentle touch of the file should correct the trouble.

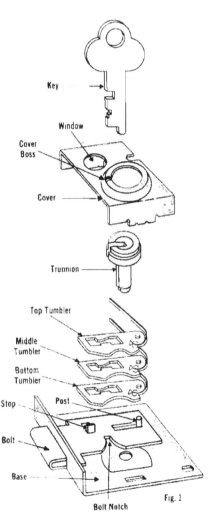

Fig. 1

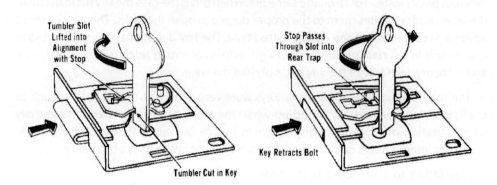

In filing keys be very careful to get the notch for each lever directly under that lever. Do not make it any wider than is absolutely necessary and always use a file thinner than the lever which is to fit into the notch. Otherwise there may be a danger of having more than one lever fit into one notch. Duplicate keys for lever locks should always be cut on a key machine, not by hand. Some levers are so thin that the slightest error in width of a cut may let an extra lever into the hole causing the lock to fail to operate, perhaps not immedi¬ately, but sometime in the future.

A key for a lever lock should be of steel for best results. Never fashion a key from a metal softer than that in the levers. Since most levers are made of brass or bronze, any harder metal will do. But avoid soft blanks. They break easily and wear quickly.

Since lever-tumbler locks are in such gen¬eral use their repair is an art that every locksmith quickly learns. The main cause of trouble is broken or dis¬connected lever springs. Whether the steel actually breaks or falls out of its slot the lock is almost certain to be jammed. Lever lock springs are ordinarily made of fine flat spring steel, slightly narrower than the thickness of the lever to which it is attached.

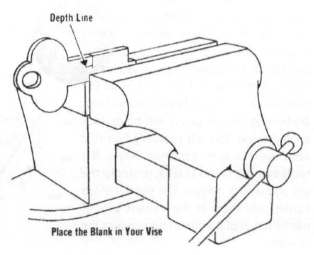

Place the Blank in Your Vise

When installing new springs re¬move from the lock case all the broken pieces of the old one. Put the levers back in exactly the same position they were in or the key will not work. In this connection, pay no attention to the numbers on the levers. Ordinarily they do not tell the position of the levers in the lock, but merely indicate the relative depth of a cut in the key. In a five lever lock, you might find levers stamped, in the order of their removal 1, 2, 1, 5, 3. Those numbers would indicate a shallow cut, a deeper cut, another shallow one, a very deep one and a medium depth cut.

The only way to keep the levers in pro¬per position is to take them out one at a time, pile them on the bench in a line from left to right and reverse the routine for replacing them.

Spring stock for replacements should not be heavier than the original, lest it cause unneces¬sary wear on the key. Do not try to use the old spring if only the end is broken off. If it is cry¬stallised in one place it will soon break at a new point. The spring itself should be cut to length and bent to approximate shape before being set into the spring slot. A spring should fit snugly into this slot, and be held in by slightly rolling the metal on both sides of the lever up over the sides of the spring. This is best done with a cross pein hammer. Do not make the lever bulge or it will bind against other levers.

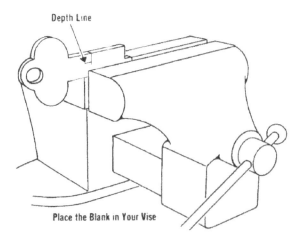

Another type of problem is worn or chewed up levers. Sometimes a steel key in use for a long time will gradually remove enough metal from the bearing surface of a lever to interfere with operation of the lock. When this happens it is best not to widen the lever gate with a file, as is commonly done, because that destroys a great measure of the lock's security. The best remedy is probably to clean up the rough spot with a fine file and then pound the lever a little thinner at the bearing surface. This will usually flatten it enough to widen the lever where required.

Sometimes the post on which the levers are pivoted comes loose. This is easily fixed with a light hammer, by riveting the fixed end into the case solidly. Occasionally there may be a broken part that will require mending. Solder is not suitable material for repair work if there is to be any direct strain or scraping against it. The metal is just too soft to be prac¬tical. It is best in such cases to resort to brazing with brass. The cost, if you do not have your own equipment, should not be high if you tell the welder before he does the job that you will dress off the extra metal yourself.

Probably the commonest complaint of the lever lock is the bent bolt. Where a lock has been jimmied, pried open or pushed against when locked, it is usually only necessary to take it apart and straighten the bent parts. Since bolts are usually made of brass or bronze they bend readily and are easily straightened. The only caution is to do it carefully and slowly, warming the metal slightly in cold weather to prevent its crystallising and breaking.

Lever type mortise locks, when they have been attacked by burglars, usually come to the shop with a cracked or broken case, especially if that part is of cast iron. While a new case is sometimes preferable, it may be necessary to have them welded if the break is not too bad. When this is done all the small parts must be removed from the lock, since intense heat will take the temper out of springs and soften brass castings. The actual welding should, where possible, be done on the outside of the case to avoid the necessity of dressing off excess metal. In many cases trouble with lever locks is directly traceable to poorly fitted or worn out keys. In some instances a new key cut a little shallower on a slightly heavier blank will rem¬edy the problem.

To pick a lever tumbler lock takes at least two tools. One is the tension device. These vary from a simple wrench to counter weighted lever arm. The purpose is to apply a gentle pressure to the bolt. The second tool is the pick used to lift the lever tumblers. Each lever is carefully lifted until its slot catches on the bolt's post.

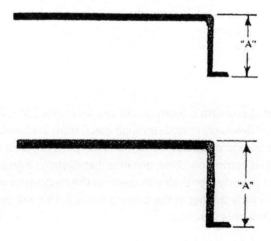

Due to imperfections in manufacturing the levers the bolt will contact one before the others. By lifting on the lever that has the most resistance it will catch on the bolt as the post enters its slot by the same amount of imperfection that allowed the bolt to contact it first. This traps the lever in the slot. Then another lever is lifted until it is trapped and then another.   Another method that requires more sophisticated tools is to lift all the levers, then apply heavy pressure to the bolt sufficient to stretch the levers first contacted and hold all the levers in place. The pressure on the bolt is very gradually released dropping each lever and catching it on the post. This method does not always work but it has the benefit of being fast and simple, requiring less skill than picking by hand.   The more precision and smooth the parts of the lock the more difficult it is to pick. In a perfect lock the levers could not be trapped and the lock would be virtually impossible to pick. But nothing is perfect and it is the imperfection of manufacturing that allows locks to be picked. It is also important that the post and the slots in the levers have crisp square corners. A round post can be easily wiggled into the lever slots and lift the levers by force of the bolt even if the levers are misaligned.

One method that was used to make lever tumbler locks more difficult to pick was a series of teeth, like saw teeth on the end of the levers and on the matching face of the post. When the bolt is pressed against the levers it engages the teeth so that the levers cannot be lifted without backing off on the bolt. A constant pressure cannot be applied and every time the bolt is backed off there is a good chance a trapped lever will be dropped.

It has also been found that it is best if the bit cuts in all the tumblers are the same shape and rest at the same height. This makes it much more difficult to make a false key by trial and error fitting. When the bottoms of the tumblers are different heights they give away the needed shape of the key.   A well made custom lever tumbler lock that combines warding and keyhole protection, along with the increased security of a system of wards, would be a very good lock in the modern world.

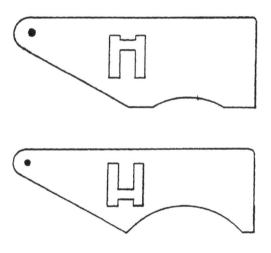

To open the lever lock introduces a new class of technique, individually lifting the tumblers to their proper height. The tension wrench used for lever locks varies depending on the type of lock. In some locks, just the key end is used to move the bolt once it has been unlocked, but there are models where the nose has a cam attached to its back end that directly acts on the bolt. In these cases a wrench that will move the nose, yet allow effective pick access, is sufficient. In the locks where the key itself moves the bolt, these are the most common, a special tension wrench is required. These Z shaped wrenches must be sized to the length of the nose. If dimension A corresponds to the length of the key from cylinder cut to tip, then that is the right size wrench.

To allow room for pick manipulation outside the lock, the A dimension should be no greater than that. Commercial picks usually come in four or five A dimensions and it is preferable buy a set rather than try making one. If you want a hand made tool, bend the profile from .040" music wire and grind each side flat for lock clearance.

Grinding the wrench profile from a flat piece of stock is also possible but time consuming. Either buy flat stock at an industrial supply shop or from a hardware store that sells cabinet scrapers, (a piece of tempered steel about the size of a playing card). Grind the wrench profile from it. For lifter tools your usual set that works disc and pin tumblers will also work the levers.

The theory of lever tumbler picking involves exerting unlocking tension on the bolt, which in turn will cause the stump to bear against the inner edge of the locked position tumbler cut-out. Once tension is applied, the specialist inserts uses the lifter to raise an individual tumbler until its gate is lined up with the stump. Since stumps are usually light parts that can be bent off perpendicular with extreme bolt pressure, the tumbler to start lifting initially is the one at the back of the lock. If the stump is bent it will contact this tumbler, the heaviest, and the others hardly or not at all.

When the stump and gate align the stump will enter into the gate slightly and catch on the stump. If tension is not released this slight entrance (only a few thousandth of an inch) will serve to hold the tumbler in the unlocked position, even if the lifting pressure is removed. The feel of the tumbler gate aligning with the stump may be quite noticeable, and may even be slightly audible. Often, the tension wrench will signal this event by

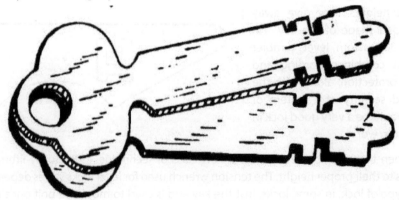

jumping slightly. The entrance or aligning may also be felt as a slight lessening in lifting resistance, which will immediately increase if the tumbler is overlifted.

Do not over lift any tumbler or you will have to start again. Some levers are provided with a special tumbler that detects over lifting and immediately locks the bolt from further movement, so be careful. Once the first tumbler has been lifted to the unlocked position proceed to the next tumbler, working from the back to the front. This tension may be slightly decreased at this point, but not too much. The feel of subsequent tumblers entering their gates will be progressively less than the first tumbler, again owing to the bending angle of the stump. Eventually all of the tumblers are lined up and the bolt moves, unlocking.

A lever tumbler may have one tumbler cut as deep as possible, adjacent to one cut as shallow as possible, owing to its unique design. This is not possible with disc or pin tumblers. It is good security because for the pick to raise one tumbler sufficiently high without touching, and possibly misaligning an adjacent tumbler, the lifter must have a high hook to put as much space between working tip and shank.

A high-low-high¬-low-high combination would be very difficult indeed, just as it is with pin tumbler locks. Over lifting is fatal. The only way to get a tumbler back down is to release tension until it falls, and other picked tumblers may also fall in the process, which can be very discouraging. Conversely, if a lever happens to drop back down while you are working it, as the pick leaves it, immediately go to the back of the lock and test each tumbler for alignment. The tumblers must be picked in order since they bind most strongly in the usual back to front order.

Occasionally you may encounter a lock where the third tumbler will not budge when you attempt to lift it. Try backing off the tension a little. If still no go, it is certain you have a lock with an uneven tumbler line and not all of the tumblers edges equally in line with the stump. If this happens, you can take advantage of this lack of manufacturing tolerance. If you apply a medium tension to the bolt and feel to see which of the tumblers is the most bound or the hardest to lift, that tumbler will be the first and the easiest to bypass, since none of the other tumblers are preventing the stump from moving as much as the bound tumbler. In other words play exists between the looser tumblers and the stump. Therefore the stump is free to advance slightly, once the tumbler that is blocking it is picked, the most bound tumbler and so on.

This is the second system of gating height, where the tumbler is cut away at the bottom to provide gate height differences. By probing the lock, and feeling for the relative amount of curvature on each lever, it is possible to get a mental picture of how far each lever must be lifted to align the gate and stump. You may need a special lifter pick with a very slim tip to feel between adjacent tumblers, but this can be a very worthwhile technique.

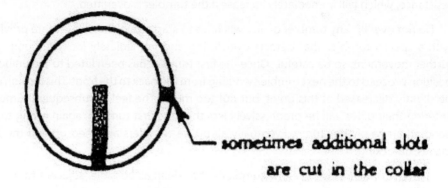

sometimes additional slots
are cut in the collar

Some lever chest and luggage locks present real problems in opening because of the small keyway and therefore limited access. For these locks have a good assortment of luggage keys and a file for on the spot alteration. Hopefully one will find the keyway, can be impressioned and a blank filed (by means of a coating of soot) to operate the lock. The levers must then be lifted as usual.

Probing the lock for a part that seems to be spring loaded will offer clues as to the type of lever and number. Often a length of music wire bent to an approximate Z configuration will serve as a good tension wrench and another length suitably bent will serve as a lifter. Try filing away the warded sides of a key to allow lifter pick access. Always suspect that what appears to be a small lever lock may in fact be a warded lock with a spring retainer that must be lifted before the bolt can be moved. The amount of lifting for a retainer is not critical, for a lever lock it would be.

KEYMAKING AND PRACTICE LOCKS

A relatively cheap deadbolt is ideal for your first practice lock. Cheap locks are good because they do not usually have any countermeasures. A deadbolt is preferable because door knob locks have a other mechanisms that allow the handle on the inside to turn when the outside is locked, whereas deadbolt mechanisms are simple and straightforward. They also provide an extremely satisfying clunk when they open. An unbranded deadbolt would make a good practice lock.

Before you do anything with your new practice lock take a close look at how it works. Note how the slots in the keyway prevent the tumblers from falling too far down. If possible remove the tumblers and observe their shapes. Be careful not to forget the order they came out in. You can even line them up on the table above the key to see how the gaps line up when the key is inserted. Also notice how the cylinders interface with the latch mechanism.

It is difficult to work with a loose lock so mount it before you start practicing on it. Once you have mastered the lock, remove it and buy a more challenging one to install.

Practice is critical. Keep practicing and you will be surprised how quickly you will master opening cheap deadbolts. This is extremely useful because cheap deadbolts are well used. It is amazing how many people trust them as they are probably the easiest tumbler locks to defeat.

Keymaking

The famous Yale key is a cylinder key. Other noted manufacturers of cylinder keys are Russwin, Corbin, Sargent, Ilco, Keil, Segal, Schlage, and Taylor, as well as many own brands and cheaper Chinese imports.

Do not allow the many names of cylinder keys to worry you. You do not have to remember them all. You merely have to learn how to tell them apart according to their shapes and grooves. As you proceed you will find the identification of key blanks is not that complicated.

Cylinder keys are made out of either brass or nickel silver. You can tell them apart as the brass keys have a yellow colour, while the nickel silver keys have a silvery colour.

Our first problem is to select a key blank that we can file down to match the cylinder key, which we will call the SAMPLE KEY. A key blank is nothing more than a key without any notches or cuts. It does not matter whose name is on the bow because our only interest is the shape of the grooves. The key blank grooves must correspond exactly to the shape of the grooves of the sample key.

The next step is to compare the grooves or keyways. This is done by turning the blanks endwise so that you can compare them at the tips. This step is very important because the grooves of a key blank must match perfectly the grooves of the sample key before it can enter the lock.

When you have finished comparing the tips, separate the pair and check the tops of the grooves near the shoulders of the keys. Compare BOTH sides. Some key blanks are so similar the only way you can tell them apart is by comparing the shapes of the tops of the grooves.

Continue to compare all the key blanks with the sample key until you find the correct blank. Then mix up the blanks in the envelope and try again. Repeat this four or five times until you are sure that you can find the blank to match the sample key.

There are thousands of different types of key blanks. To attempt to remember all of them would be impractical. Key blank manufacturers publish excellent catalogues which list all of them. A key blank catalogue is a necessary tool for the locksmith. It helps keep stock in order and also how to identify the blank used by name and number. A good key blank catalogue also lists comparative numbers so that a locksmith can find any key blank, regardless of the brand name that appears on the bow of the blank. Most key blank manufacturers make key blanks to fit any lock. That is why you will often find that keys will fit into various makes of lock. So long as the blanks are correctly grooved and sized, they will enter the lock intended for them.

Loosen the vice jaws and again raise the key up about 1/16 of an inch and squeeze again. Repeat the process once or twice more, raising the key each time. As you do so, you will be pulling the strip around the post until you create a snug fit.

The strip has now become a snug mask over the bit and post of the sample key. This does not mean it will not move if you pull it or try to slide it off the sample key. Now move the mask from your sample key.

[[IMAGE FILING KEY WARDS]]

The simplest method of darkening the bit is by smoking it. You must darken all the tumbler cuts, ward cuts and edges of the bit completely. Light a match and hold the key over the yellow part of the flame. Use a pair of pliers when holding the key because it will get hot. Only let a thin coat of soot form on the bit. You are only trying to discolour the bit enough to provide a dark background. If you smoke the bit excessively the loose soot will smear or flake off.

Replace the mask on the key with the long flap in front. Trim the mask with your warding file until it becomes an exact outline of the bit of the sample key. Place the sample key and mask in your vice.

In this position you will be able to file on the cuts without difficulty. Do not change the position of the sample key and mask once they have been clamped in the vice. Leave them where you put them or they may shift as you loosen the vice and try to reposition them. This would result in a very inaccurate outline.

Use the wide cutting face of your file and file the mask to the exact overall size of the bit of the sample key. File slowly and carefully. Do not angle or shake as you make your strokes. Keep the cutting face of the file as parallel as you can to the edge of the sample key's bit. File as close to it as you possibly can without cutting into it. Watch closely as you file so you can stop filing as soon as you begin to scrape the soot from the bit of the sample key. Use light forward strokes so as not to bend the mask.

A file will only cut on a forward stroke. File slowly using long and steady strokes. After each forward stroke lift the file slightly and replace it on the mask ready for your next stroke. Never drag a file back across a surface after you have completed the forward

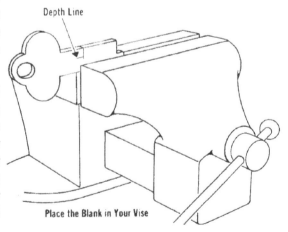

Depth Line

Place the Blank in Your Vise

stroke. A backward stroke will bend the mask away from the bit and make filing difficult.

With the mask trimmed to the square outline of the bit, you may now file forward tumbler cuts. Use the edge of your file using long and steady forward strokes only.

Before you remove your sample key and mask from the vice, study them very carefully. Look at them from all angles, front, rear, left edge and right edge.

- Are all cuts full and square?
- Is the pattern even with the edge of the bit?
- Are there any burrs?

Have you filed your mask so that it is a faithful silhouette of the sample key? If not, trim where necessary. When you are satisfied that the mask is correct, remove both the mask and sample key from your vice.

Push the pattern over the key blank right up to the shoulder place the blank in the vice. File the blank so that it is an exact replica of the pattern. Proceed carefully and avoid cutting too deep.

Remove the key from the vice and check it carefully. If your cuts are square and full, remove the pattern and place both items aside.

Barrel keys are duplicated in very much the same way as bit keys. However barrel keys do not usually have a shoulder. Therefore there is nowhere on which your mask may rest in order to maintain its proper position on the bit. If the location is not fixed the mask could easily slip while filing. For this reason, while forming the soft metal strip that will become your mask, leave a small piece of metal behind the bit that may be folded down on to the shank. With this lug pressed up against the rear of the bit you will ensure the mask remains properly aligned throughout the filing process.

When clamping a barrel key in the vive, it is always necessary to insert a metal or wooden plug in the hole to prevent the vice from distorting or crushing it.

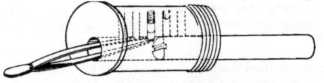

First Step in Setting up a Cylinder

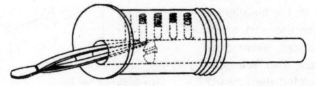

Second Step in Setting up a Cylinder

Ready Made Keys

Very often you will find that certain types of bit keys and barrel keys may be obtained ready made from the factory. It is better to order ready made keys from your supplier and have them in stock instead of the blanks. These ready made keys are so common that you will get calls for them time and time again.

Bit keys can vary in design from very simple to very complex. One design feature that occasionally appears on bit keys is a side groove. The purpose for such a groove is to restrict entrance into the lock by any key other than the one containing the proper groove. Whenever there is a side groove in a sample key, you must file a corresponding groove in the duplicate.

To locate the position of the groove use a groove guide. It is another mask that serves two purposes. It shows you where to file the side groove in the bit and it helps you measure the depth of the groove.

This is how a groove guide is made. Form a soft metal strip into a mask. This time bend the strip so that both flaps are approximately the same size. This is done because some bit keys have grooves on both sides. If yours does, you will be able to locate both grooves and measure their depths using only a single mask.

After you have pressed the strip snugly around the sample key, remove the assembly from your vice. Next slip the guide down along the post until you can see the groove. Using a scriber now make a scratch mark on the mask indicating the location of the top and bottom of the groove. Then remove the groove guide from the sample key and slide it on to the post of your duplicate. Mark both edges of the bit. If these marks do not stand out clearly try smoking the bit of the blank first. Take a straight edge and connect the scratch marks by scribing two lines as shown. These lines would represent the width of the groove.

Having marked the location of the groove, determine how deep it must be filed so that it will be equal in depth to the groove in the sample key. Stand one flap of your groove guide inside one of the walls of the groove in the sample key. Scribe a line on to the groove guide where it reaches the top of the groove. This now becomes your temporary depth guide.

Complete your key by cutting the groove into the blank using a warding file. Check your work from time to time to determine whether or not you are filing wide, deep and evenly enough. The purpose of the groove is to enable the key to pass an obstruction in the keyway. A groove which may be slightly wider or deeper than the one in the sample key is acceptable provided you do not weaken the key.

Do not attempt to begin filing the groove with your warding file lying flat on the bit of the key. Tilt the back end of your file up slightly so that you will begin to file the far end of the groove first. After several strokes you will have filed enough to enable the shallow beginning of the groove's walls to help guide your file. Lower the angle of your file placing it flat on the bit and continue to file the groove. Following this procedure will ensure avoiding the having your file drift beyond the required width of the groove. The temporary depth guide is usually all you need to determine the depth of the groove because the depth is not very critical on most bit keys. The groove just has to be deep and wide enough to pass into the keyhole. Some locksmiths prefer to use a slot gauge. This is nothing more than a piece of soft metal with a slot filed in it equal to the exact thickness of the bit at the bottom of the groove.

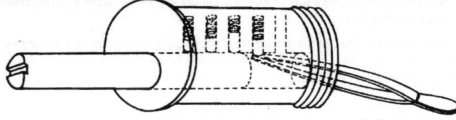

Third Step in Setting up a Cylinder

Checking Groove Depth of Duplicate

When comparing your sample and blank key for thickness of bit, make sure the blank you choose is not too thin. If you attempt to file a groove in a blank which is too thin the strength of the bit may be severely weakened. The bit may possibly bend or break at the groove when you try the key in the lock.

A Tip on How to Measure Post Diameters

Many locksmiths who duplicate large numbers of bit keys find it easy to use a caliper or drill guage to measure post size. These instruments may be used in the following manner:

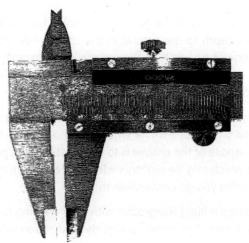

Remove the key. Be careful not to alter the setting.

Attempt to slide the post of your blank key into the opening. A proper diameter post will slide into the opening just barely dragging on the measuring surface as it moves. A post which is too large will not enter without forcing the opening wider. A post which is too thin will enter easily but wobble within the jaws of your caliper.

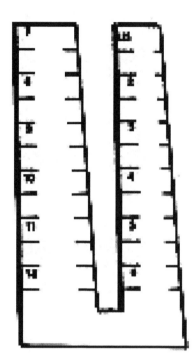

- Attempt to fit the post of the sample key into the holes in the drill gauge.

- Note the number of the hole in which the post fits best.

- Remove the sample key from the gauge.

Try to insert the post of the blank key in the

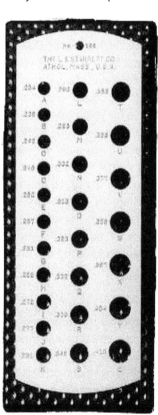

same hole. If the post fits as snugly as the post of the sample key you may use this blank to form your duplicate, provided the other dimensions of height, width, length and thickness are acceptable. You should not use a blank whose post was too large to fit into the same hole as the sample.

If the post of the blank key enters the hole but fits loosely, find the hole in which it fits best. Note the difference in dimension between the two holes. If your blank is no more than 1/16" or .060" smaller than the sample it may still be used.

Today's locksmith is a specialist in security. He has to know about different types of locks and how they are applied in order to provide the best protection. He must have an overall knowledge of the other types of security devices that add to the protection of any given premises.

You need to know which locks should be installed in what type premises, so that your customer can derive the greatest security benefit from your knowledge.

Residential Protection

Most people will not call a locksmith until they need help. Your customer's home may have been burgled. The forced entry may have left his door lock useless. Maybe the door lock is inoperable and your customer is afraid the burglar may come back. Maybe he wants a second lock installed on the door to increase security. In any case a residential customer who requires your services probably fears for the safety of his loved ones and the security of his possessions. This is an opportunity to perform a service for your customer. While you are at his home doing the work he has requested of you, offer him your services to inspect all the other protective devices that guard his home. You can make him aware of other vulnerable security points. In a private residence internal doors separate the rooms. Entrance doors permit access to the residence from outside. In burglar proofing a residence we are not concerned with internal doors, except where they provide access to sensitive storage areas. Many entrance doors have more than one lock. The lock installed by the builder of the residence is the primary lock. It may be either mortise or cylindrical. The locks installed afterwards to increase security are auxiliary locks, and they may be either tubular or surface mounted.

Primary locks

There are three basic primary lock functions that can be found on an entrance door. A simple latch provides the ultimate in convenience for the resident. When leaving the premises he closes the door behind him and it locks. A deadlatch operates the same way, but once the door is closed the latch bolt is secured in the locked position and acts like a deadbolt. A deadlock uses a separate bolt that must be thrown after the door is closed. This provides the greatest protection, but is much less convenient.

Let us examine these three functions more carefully.

Latches

A latch has a bevelled bolt that is spring loaded to enter the strike when the door is closed. This automatically secures the door. Turning the inside knob will retract the bevelled bolt, as will turning the key in the outside cylinder. The main advantage of a latch is convenience. The disadvantage of a latch is since a latch can be easily breached, often without detection. It is common burglary practice to force back the latch by inserting a plastic credit card or a thin screwdriver between the door and frame. If the door-to-frame tolerances are large enough the burglar can accomplish entry, without leaving any signs of forcible entry. The technique is called loiding. With the latch bolt depressed the door can be opened. Latches should not be used as the primary lock on an entrance door.

Deadlatches

The convenience of a latch is the attraction to this design. A deadlatch offers convenience too, but with more security. This is how it works. At the edge of the door you will see a bevelled bolt, characteristic of a regular latch. If it is a cylindrical-design deadlatch a small button will project from the lock and rest alongside the bolt. In a mortise design the deadlatch device will take the shape of another bevelled bolt, slightly smaller than the latch bolt and placed higher up on the lock body. In both designs, the latchbolt enters the strike to secure the door after it has closed.

The deadlatch device does not enter the strike, but is depressed into the lock when the door is closed. While depressed the deadlatch acts as a stop and secures the latch bolt in the extended position. The bolt can be operated by turning the inside knob or using the proper key in the outside cylinder. The deadlatch cannot be levered open. It can be jimmied open. This is another common burglary practice in which the burglar forces apart the door and frame to release the lock.

Deadlocks

A deadlock provides the greatest degree of primary lock protection. It uses both a bevelled latchbolt and a rectangular deadbolt. When the door is closed the latch secures it. At this point turning the inside thumb-piece or using the proper key in the outside cylinder will cause the deadbolt to enter the strike. A deadlock effectively double locks the door with two bolts, a deadbolt and a

latchbolt. Although even the secure deadlock is vulnerable to jimmying there are ways of preventing this. Auxiliary locks provide part of the answer.

Auxillary Locks

There are many locks classified as auxiliary locks. The most common is the rim, or surface-mounted lock.

Rim Night Latch:

Three basic styles of the rim lock are available to meet most auxiliary lock requirements. The simple night latch, the least secure of the three, works exactly as the primary latch. It provides convenience in that the resident slams the door shut behind him and it is locked.

Rim Deadlatch:

A rim deadlatch operates differently from a primary deadlatch, although they carry the same name. Both types can secure the latch bolt in the extended position for added security against loiding. In a rim lock this action is not automatic. Once the latch has entered the strike, a button on the lock body is moved to set the latch to deadlock. Since this must be done from inside the premises, this extra measure of security only applies when the resident is home.

Rim Deadlock

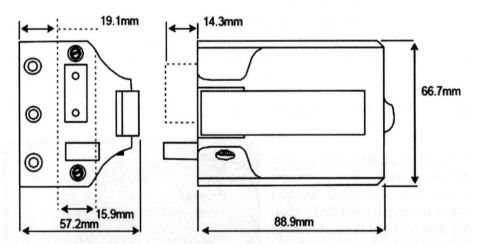

A more secure rim lock is the deadlock. It is installed the same way as the rim latch and deadlatch. This design uses a bolt that must be operated each time the door is opened and closed. In most styles a heavy rectangular bolt is used. Loiding is out of the question but the door can be jimmied open. A long steel bar is inserted between the door and the jamb. The length of the bar provides leverage that forces apart the door and jamb sufficiently to permit the bolt to come out of the strike.

Lock manufacturers have responded to jimmying by designing longer deadbolts that are less likely to come out of the strike. Other designs, which they call jimmy-proof, are also available. One popular jimmy-proof lock uses a bolt with two barrel-shaped projections that drop vertically into the strike. Spreading the door and jamb apart with a steel bar will not cause this bolt to come out of the strike.

Tubular Deadlocks

A major fault with many rim locks is that they are ugly. In a residence where your customer is understandably concerned with appearance, he may find the rim lock especially objectionable. The answer to this problem is the tubular deadlock, which works exactly like the rim deadlock, but is installed inside the door.

If a burglar breaks the glass or panel he can release a conventional lock by reaching inside and turning the thumb piece. Double cylinder locks prevent this. Tubular deadlocks can be purchased with either standard or long-throw bolts. Where a door fits poorly in a jamb or where an extra measure of security is required, use the long-throw bolt which measures at least one inch.

Brace Locks

Brace locks provide the greatest degree of residential door protection available. In this design the cylinder and lock body are installed like the rim lock. You will not find a bolt in a brace lock, so no strike is used. Instead, the lock operates the brace, a long steel bar that anchors in the floor. In the locked position the top end of the bar is cradled in the lock body. Pressure on the door forces the bar more securely into the lock and the anchor plate in the floor. To open the door the bar is moved to the open side of the lock where its top end is freed. As the door swings open the bar rides out of the lock and is kept from falling by a U shaped support. The resident steps inside, lifts the bar from the floor anchor, slides it out of the support and places it aside.

The brace lock is not meant to be convenient or good looking. However, it is the most secure residential design yet developed. It is used extensively in high crime areas where customers are willing to sacrifice convenience and beauty for security. The one physical limitation of the brace lock is that it can only be used on a door that opens into the premises.

Chain Locks

 Police constantly caution people against opening their door to strangers. A chain lock permits the resident to open his door sufficiently to see and speak to persons outside, but will not allow the door to open fully.   The slide is mounted to the door and the lock support is mounted to the jamb. One end of the chain contains the lock mechanism, and this is inserted into the support. The other end inserts in the slide.

In this position, the door can only be opened to the limit of the slack in the chain. When the resident wants the door open completely, he closes the door, slips the chain out of the slide and then reopens the door.

When the resident leaves home, he secures the chain lock from outside. He leaves the slide end of the chain engaged, but uses his key to unlock the support end. Just before closing the door, he reaches in, drops the lock mechanism into the support, and removes the key. The lock is now secured, and he is on the outside.

The chain lock was never intended to take the place of an auxiliary lock. It is an accessory item designed to let the resident screen callers. Since the chain lock can be secured from outside it adds another measure of security. But if a burglar succeeds in opening a door sufficiently for the chain lock to be challenged it will no stop him gaining entry. He can cut it or just push the door and rip the device off the door frame.

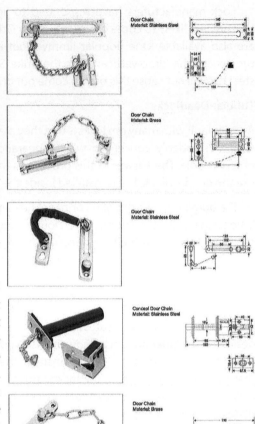

Sliding Glass Door locks

Sliding glass doors can be found in many newer homes. Most are in the rear of the house making ideal as points of illegal entry. Unfortunately many homeowners secure sliding glass doors using only the simple cam-latching device built-in by the manufacturer. This is poor security. Many locks have been designed to secure sliding glass doors. These vary in design to be used in cases where only one or both sections are movable. Some locks are simple jamming mechanisms, others are key operated.

A bar lock is the easiest way to protect a one section sliding glass door. The bar is hinged at the far end of the opening and pivots down to a support mounted on the door. In this position the door cannot open. When lifted out of the support the bar can be pivoted up where the hinge bracket will grasp it. This frees the door so that it can be opened.

A two section set-up can be protected by using a specially designed key lock. One lock protects both doors since it secures one to the other so that neither can slide. This is a most secure set-up for three reasons:-

1. The key lock makes the doors secure, even from the inside.

2. When the lock is installed it is out of view from outside.

3. With the doors locked to each other, side-to-side play can be minimised.

The lock is attached to the inside track door and a hole is drilled in the outside track door to receive the bolt. When the lock button is depressed the bolt enters the hole in the outside door, effectively securing the doors. A key withdraws the bolt when the resident wishes to move either door.

Window Locks

If you are burglar-proofing premises, do not neglect the windows. You can be certain a burglar's first attempt will be made through an unprotected window, because that is the easy way. Most windows in a residence are the double-hung type. A cam-locking device is used, mostly to prevent the upper window from closing down due to gravity. Such devices also prevent either window from being forced open from outside. But the devices are not locks, since they are not key operated and are not secure.

The burglar will usually exit through a window. But if the windows are equipped with key operated locks they cannot be used for this purpose. The only alternative is to break out of the premises by forcing a door or shattering a window. This means noise, and noise often leads to the burglar's apprehension.

Commercial Protection

Lock specification pertaining to residences also applies to commercial establishments. But businesses demand even greater security. There are many additional products that will provide the necessary added protection.

High-Security Pick Resistant Cylinders

Conventional pin-tumbler cylinders that contain five or six sets of pins are supplied with most locks used in commercial applications. But in cases where your customer may be concerned with unauthorised persons making duplicate keys, or if he just wants a cylinder that is difficult to pick open, a high-security cylinder is in order. A true high-security cylinder is of a sophisticated design and uses keys that restrict unauthorised duplication.

High-security cylinders are not new to locksmithing. Several designs have been around for years, each expressing the individual manufacturer's answer to the age old problem of securing a person's valuables.

Medeco Cylinders

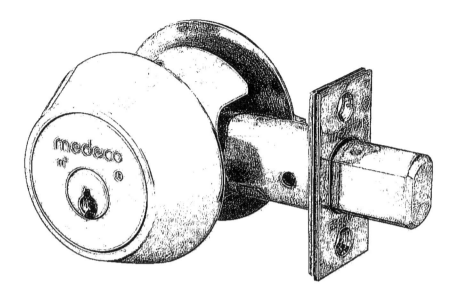

To the Medeco Security Lock Company, manufacturers of Medeco cylinders, high-security means using specially designed pin tumblers with angled bottoms and keys with cuts set at corresponding angles. Before the key can turn each pin must be lifted and twisted into proper position. Three thin rods and a disc made of hardened steel help prevent drilling. Medeco keys cannot be cut on conventional key machines. Specially designed machinery is required which is available only to bona-fide locksmiths and suppliers. This policy, coupled with the unconventional design of the Medeco key, restricts unauthorised key duplication.

Cylinder-Guard Plates

Not every burglar is adept at lock picking or has access to keys which he can copy and later use to enter premises. Most burglars rely on brute force to gain access.

One popular method of forcible entry is to wrench or pull the cylinder out of the lock and then operate the lock manually to the open position. Another technique involves drilling the cylinder to make it ineffective. You can protect your customer from forced entry of this kind by a cylinder-guard plate.

A cylinder-guard plate shrouds the face of the cylinder with an impenetrable steel cover. Only a small hole in the plate permits the customer to insert his key and operate the lock. The plate is held by bolts that pass through the door from outside with nuts and lock washers on the inside. The set-up is not pretty but is very effective.

It is not uncommon for a locksmith to compromise with his customer on a form of protection he knows is best if the customer objects on the basis of appearance. Hardened-steel cylinder rings provide just such a compromise between leaving a cylinder unprotected, and installing a guard plate. The ring turns idly around the cylinder if any attempt is made to unscrew the cylinder from the lock. It cannot be pried off the cylinder because it is hardened. The cylinder ring still does not measure up to a guard plate because it cannot protect against drilling the cylinder or pulling it from the lock.

Latch Guard Plates

If a burglar cannot gain access by drilling, wrenching or pulling the cylinder, he will try to jimmy the door open and mutilate the door. With a secure deadlock, preferably a jimmy-proof deadlock, there is a good chance that the burglar will not get in. But the damage he causes may be costly to repair.

Latch guards cover the opening between the door and jamb at the lock bolt. They are held to the door by two bolts inserted from outside, fastened by nuts and lock washers on the inside. With the door a latch guard provides complete protection against jimmying and loiding without any modification to the lock or its installation. Just the presence of a latch guard on a door is enough to change the minds of some burglars looking for an easy hit.

Pushbutton Locks

A commercial establishment can be an office with any number of employees. It could be a men's clothing shop solely run by the owner, a large supermarket, a shoe store, stationery or pet store. These are all different businesses, each with its own peculiar needs. The pushbutton lock solves the problem of too many keys in circulation. But what if many people must have access to a locked area? The answer is a lock that operates without keys, such as the pushbutton design.

Pushbutton locks operate a bolt to secure the door and in this way are similar to key-operated locks. Instead of a cylinder outside there is a panel with numbered or lettered pushbuttons. To operate the lock the buttons must be pushed in the right combination. This permits the bolt to withdraw and the door opens. To lock the door the bolt is thrown by a small lever.

How secure are pushbutton locks? They are secure against picking since no cylinder is used. Physically they resist forcible entry as well as any key-operated lock. The number of possible combinations far exceeds the number of key combinations you would expect in even the best cylinders. These facts present a strong case for the pushbutton lock.

Office Equipment Locks

Offices use the latest in time saving equipment. Laptops, for example, are commonplace. Unfortunately they are popular with burglars who can dispose of expensive goods quickly at a fraction of their true value. There can be good locks on the doors to an office and high-security cylinders installed to operate them. If a burglar makes his way through these many valuable items of equipment may be taken. How to stop the office burglar? Individually lock each piece of equipment so it cannot be removed from the premises. This is done easily with locks specially designed for this purpose. A good example is the Bolen 200 Maximum-Security lock. In this design the equipment is anchored to its desk by a hardened steel bolt that passes through the lock base and threads into the chassis. When the pick-resistant lock mechanism is inserted in the base and secured, access to the bolt is sealed. The equipment is secure.

A single lock is sufficient to protect most office equipment in moderate risk crime areas. If a customer needs further protection, however, install a hardened-steel locking bar with the Bolen lock. This provides two-point security over distances up to twelve inches using a single lock! To secure large equipment such as monitors or copying machines, where the distance between two locking points is more than twelve inches, two Bolen locks are recommended for top security.

Filing Cabinet Bars

A business can store some records in a safe but it is not practical to keep all of them this well protected. Filing cabinets are used for the less important, such as records, but sometimes these require a measure of security not available in the average filing cabinet.

You can install filing cabinet bars that make a simple filing cabinet almost as secure as the safe or vault. Installation is simple. The bars are sold in kits containing all the necessary hardware and instructions. Bars are available in different sizes to fit almost every type of filing cabinet.

Door and Window Gates

Large plate glass windows form the fronts of many retail establishments and help make the displayed merchandise in the window saleable, but also may make it easier to brake in.

Window Gate

To the burglar, an iron gate on storefront means that he cannot get in even if he breaks the window. To thwart malicious window breakage the gate can be made of a fine mesh that will stop anything from a baseball bat to a rock. The shop is protected and the window can still be used for display. Most gates fold aside or are completely removable for daytime use.

The locksmith need not be the manufacturer of the window gates he installs. In many cases the locksmith does not handle the installation. A substantial profit can be made just by recognising the need and sub-letting the work.

Industrial Protection

Industry has its own security demands. Residential types of protection will help make an industrial establishment secure. But you may have to go several steps further in order to achieve the measure of security demanded by industrial customers.

Panic Bolts

To comply with local building codes, all industrial establishments must have the specified number of clear access emergency exits. All emergency exit doors must be equipped with locks that can be opened from inside without a key.

A panic bolt has no key operation from inside. This makes it perfect for use on emergency exits. The few seconds a person might lose fumbling for a key could make the difference between life and death in an emergency situation. A panic bolt uses a long bar that stretches the width of the door. This is another safety feature since a person might not be able to grasp a knob etc. in an emergency. A natural reaction is to push forward, and pushing the bar of a panic bolt opens the lock for a quick exit.

What does this have to do with burglar-proofing? Regardless of what type of lock your customer wants, there are certain legal requirements that you must observe. When you are called upon to install a lock on an emergency exit door, you cannot let your customer talk you into installing a high security burglar resistant lock that uses a double cylinder. You cannot even install a simple deadlock if local building codes specify that a panic bolt is to be used.

How do you provide adequate burglar protection for an emergency exit door? Panic hardware is usually constructed of top grade materials by the leading manufacturers of lock hardware. If the door is used as an exit only there is no need for a cylinder or knob on the outside. If the door is used as a combination entrance and emergency exit, a cylinder and outside knob must be used. This design can be made more secure by installing a pick-resistant cylinder that will operate the lock. A cylinder-guard plate and latch guard can be installed to complete the outside protection.

Exit Alarm Locks

To the industrial customer it can be just as important to keep people from breaking out of the building during the day, as it is to keep burglars from breaking into the building at night. Since some exit doors may be in remote areas of the industrial building, they may be concerned with people using these to leave the premises with merchandise they are not authorised to have. Pilfering can be a major concern.

Conventional locks help restrict access to areas where only certain persons are permitted to be. This is fine for storage rooms, tool cribs or other security sensitive areas where installing a deadlock is legal. As the lock on a fire or emergency exit door must be opened from inside, easily at all times and without a key, it is illegal to install a regular deadlock on such a door.

Exit alarm locks are commonly used where conventional locks cannot be used. Rather than securing the door, they permit free exit at the push of a handle. A pilferer is discovered because the instant the lock is released and a battery-operated horn in the lock sounds. The alarm can only be silenced when the lock is reset with the proper key.

The use of exit alarm locks is not limited to fire and emergency exit doors. These can be installed on any door your customer wants protected. When operated with a key, the exit alarm lock will function as any ordinary deadlock.

Electric-Release Strikes

The management of an industrial business may wish to limit access to certain areas. A simple lock might be suitable in some cases. The keys could be issued only to those who have business in the specified area. But so many keys might exist that the security of the lock could be totally defeated.

Let us assume the critical area is the office and the office door is equipped with an electric-release strike. The office receptionist sits at a window and screens the person wishing entry. If he is cleared, the receptionist presses a button to release the electric strike on the office door. He can now enter. For the would-be burglar or any other person denied entry through the specially-equipped door, the electric release strike holds fast, and the door remains secured. The electric-release strike is compatible with most locks. It is mounted in the door jamb and has a flange which traps the lock bolt when the door is secured. The flange is operated by an electric solenoid. When the circuit is completed by pressing a switch, current flows to the solenoid. The door can now be opened as the flange in the strike withdraws under the pressure of the lock bolt.

Electric-release strikes can be used in industry and also in apartments, commercial businesses and banks. The installation is simple, profitable and effective against burglars and prowlers.

Double-Bar Locks

The Fox Police Lock Company of New York designed a special double-bar lock. The lock is not pretty but is rugged, secure and is made specifically to meet industrial requirements. Other manufacturers have produced their own versions of the lock.

The double-bar lock is an auxiliary design that mounts in the centre of the door at chest height. The two steel bars that secure the door are controlled from outside by a rim cylinder and from inside by a turn knob. Each bar is supported by a brace mounted near the edge of the door that guides it into the strike. The two bars of a police lock provide two-point security. The door will stand fast against prying, jimmying and even removing the hinge. A cylinder guard plate, available with most models, protects against prying or wrenching the cylinder from the lock.

Other Security Devices

This chapter has given you a true insight into what lock security means. But it is the nature of this trade to constantly develop new and more sophisticated types of security equipment. That is what keeps the locksmith ahead in the constant battle against crime. Many types of devices are available only to you as a qualified locksmith. The do-it-yourselfer is not able to obtain the security devices that may be needed. Your customer needs you. And you need to be well informed.

MISCELLANEOUS LOCK OPENING METHODS

Lock picking must be divided into two categories which are commonly confused. The first category is picking, the act of carefully manipulating one pin at a time for the expressed purpose of duplicating the action of the proper cut key in a given cylinder, by something other than the proper cut key itself.

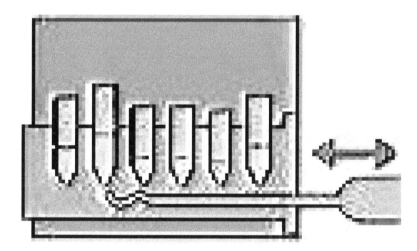

The second category is raking. Either of these techniques is intended to be a method of convenience for opening locks in emergency type situations.

The method for making keys referred to as impressioning would be far more desirable, since both processes take about the same time and only one yields both an open lock and a working key. However there are times when picking is the most logical method to use, such as when someone is locked out of a house or car and the keys are inside. Both methods are predicated on their efficiency and, should either take an undue amount of time, it is questionable how worthwhile they are when a method such as drilling is so quick and sure, though more expensive.

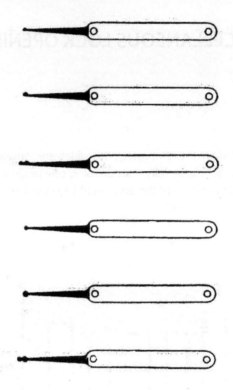

In order to understand how to compromise a lock, there are certain steps which are essential to laying a proper foundation.

They are:-

- a thorough working knowledge of the lock mechanism and how it functions
- the ability to recognise these factors so that you are able to overcome them

What is it that actually allows us to pick a lock? It is the inability of the manufacturer of any product to machine parts to an almost flawless level of tolerance. Even if they could reasonably approach their goals, the expense alone would be astronomical. Therefore as locksmiths we are able to pick a lock.

To see specifically what is involved, we must look at a typical cylinder. The tolerance inadequacies can be categorised for easy reference. The first is the difference between the plug and the shell. An acceptable amount of difference is approximately .005 or about .0025 all around the plug. The process by which the keyway is cut into the plug is called broaching. This process is easily observed when a blank or cut key is inserted in the keyway and play is felt due to a significant tolerance differential.

Probably the most significant problem of this sort is the drilling of the chambers.

This takes three forms:-

- plug diameter differential
- off-centre chambers
- concentricity

This is caused by the cost effective but necessarily imperfect process used to manufacture these cylinders, namely gang drilling. This a process by which you drill all the chambers at once and sequential drilling where you drill one chamber after the other. In either case both methods are imperfect because the drill bit itself changes a microscopic amount each time it is used to drill a chamber. It is no surprise then, that after a hundred or a thousand holes the diameter and the centring functions based on its original diameter are no longer accurate. However, in deference to the   manufacturer, he could not possibly stay in business and change the bit for each hole of set of holes. We are therefore left with a necessary evil, but one which we can use to great advantage.

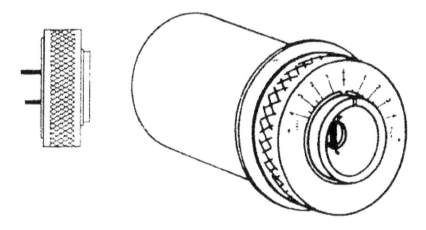

When turning tension is applied to the core, without the proper key inserted, again tolerance plays a large role in the next operation. Not all of the pins will bind at the same time. Locate the pins, lift them to the correct position (shear line). Follow by doing the same to the next pins to reach the cylinder housing. The only objects which keep the lock from opening are the pins.

In order to pick a given cylinder, firstly ascertain whether or not the cylinder can be picked. Does it operate? Can you manipulate each individual group of pins within each pin chamber?

If you can, then proceed with the picking and/or raking process.   If not, there is another alternative. Faced with the problem of frozen pins in one or more chambers, the best strategy is to clean and lubricate the lock. After the application of any solvent or lubricant, impressioning will become difficult. if not impossible.

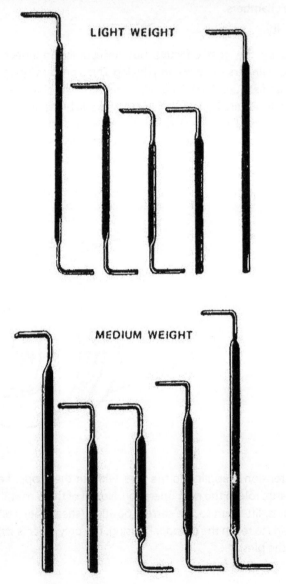

LIGHT WEIGHT

MEDIUM WEIGHT

Tool Design

Tool design is the direct result of the function it will be required to perform and falls into one of three major categories:-

- The hook tool, used when the adjacent bottom pin lengths are significantly different. This tool is advantageous for this type of situation, as it allows you to get behind the larger pins in order to properly reach the smaller ones and manipulate them open.
- The diamond pick, which is advantageous due to its design in the manipulation of wafer tumblers, which are more fragile and spaced much closer together.
- The rake is intended to do just what its name suggests, and is ideal for those situations where all the tumblers are approximately the same size or gradually rise and fall. The tools required for raking are the rake, the diamond or the ball pick and a tension tool.

All raking and picking tools are referred to as picks. Other individual styles of picks are usually just a modification of one of these groups.

The other tool used in the act of picking is the tension wrench, or more properly, the turning tool. This tool is as or more important than the pick itself, but is often overlooked. Too much pressure has defeated more would be pickers than the wrong type of pick. Only use the lightest amount of pressure necessary to turn the lock. Any more and you bind the pins so tightly that you make them work against you instead of for you.

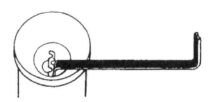

Grooved ends, rigid tension tool

Double-sided tension tool

Turning tools come in basically six groups:-

- Lie, medium and heavy duty material
- narrow medium and wide widths

Before using the tension tool try raking with the pick a few times. While inserting the pick all the way in the keyway with the tip in contact with the pins, remove the pick with a quick motion keeping an

Twisted, flexible tension tool

Various tension tools in position

upward pressure on the pins. Repeat this operation again, in slowly and out with a slight snap. Now you are ready to use the tension tool.

There are many tension tools to choose from. To start with choose a tool of medium weight and length.

The most common techniques for lock picking are:-

- raking, where a rake type tool is gently, or in some cases vigorously, pulled along all the tumblers in a rather general way
- targeting for specific individual pins, picking each individual chamber
- a combination of the previous two, you rake and then specifically target those pins you may have missed during the initial raking attempts

The pick gun is a tool that works on the principle of percussion much like cylinder rapping. It is really an effective method once you have mastered the timing necessary to make it work. It consists of the following procedure. Put the tip of the pick gun into the cylinder keyway to be picked. Make sure that the pick will strike the pins at a right angle, pull the trigger. At virtually the same moment the bottom pins are hit, the percussion causes all the top pins to fly straight up towards the top of the pin chamber for an instant, creating an enormous gap. It is in that instant that you must turn the plug with your turning tool, opening the lock. It is this ricochet effect that makes this unique tool so valuable in situations involving specialty pins and cylinders.

During the lock picking process the barest amount of turning pressure is exerted while you feel the condition of the pins in the chambers. The ideal condition is matching top and bottom pins in each chamber to maintain the same pressure in each chamber to ensure the best possible cylinder operation.

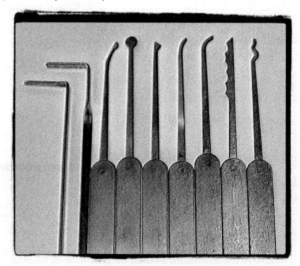

There are three conditions in which you can find the pins in any given chamber, once you have ascertained that the lock is operating properly and is therefore pickable. The pins, due to the problems with tolerance differentials acquired unavoidably during the manufacturing process, will pick only one at a time no matter how short that span of time may be. When doing your initial raking, the first condition is that the pin is in the unpicked position.

The second possibility is that the pins in the chamber are merely bound. The final possibility is that the pins in the chamber are under pressure, but not bound up. Simply continue the process of analysing the condition of each chamber until they are all picked and the lock is opened.

Note:

The pins will not necessarily pick in their regular order. This means pin No 1 will not necessarily pick first. Perhaps it will pick fifth and pin No. 3 will pick first, and so on. Raking is the most common method used today. It is the fastest to use and the quickest to learn. The raking method will work in opening most cylinders where there is not a sudden change in pin sizes, such as a combination of 7-2-6-1-8, where there is one long pin, one short pin, one long pin, etc.

The pick is able to come into contact with all pins. The pick you choose for raking should be able to move in and out freely in the upper half of the keyway so it will come into contact with all pins.

The tension tool and its use are the whole trick to raking or picking. Insert tension tool into the bottom of the keyway.

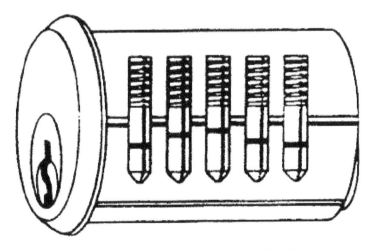

Clearance between plug and body

Then apply very light tension in the direction to unlock the lock. Do not use too much tension. You must develop a light touch with the hand that applies the tension. If tension is too heavy the top pins will bind below the shear line and will not allow the breaking point to meet the shear line.

With light tension applied, go through the raking operation, in slowly and out with a snap with upward pressure on the pins with the tip only. Repeat this operation three or four times. If the plug does not turn and open the lock, release the tension on the plug. Before releasing tension, put your ear close to the cylinder and listen for the sound of the pins clicking back into the down position. Release tension slowly so you can hear all the pins. If there is no sound you were applying too little or too much tension, not allowing the breaking-point to bind at the shear line.

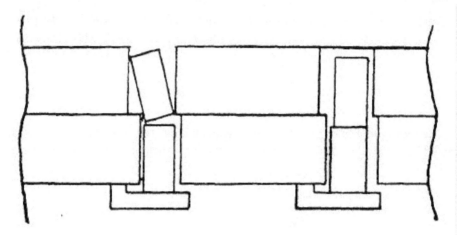

Repeat the raking operation varying the tension, lighter or heavier than on the first try. With practice you will gain the right touch in applying tension and find you can open most cylinders in a few rakings. Set up a cylinder with only a two pin combination to start with for practice. Mount the cylinder on a large mount or held firmly in a vice. Do not try holding the cylinder in your hand while raking it. Then go on to a three pin until you can rake a seven pin cylinder.

Some will open very easily, regardless of the pin combination. This is due to the poor construction of some cylinders. Generally the lower cheaper the cylinder, the easier it picks. The cheap cylinder is manufactured with greater clearances on all parts so that the cost of assembly will be kept low.

The following characteristics are commonly found in low priced cylinders:-

- too much chamfer on the top of the bottom pin
- die-cast plug and body with poor hole alignment

- oversized pin holes
- too much clearance between plug and body

Oversized pin holes leave ample pin clearance. This is an aid for the manufacturer in the assembly of the cylinder, but it is also an aid for the locksmith who must pick the cylinder. More expensive cylinders are manufactured with much less clearance. But in spite of better construction you still can pick or rake it open. It might take a little longer, needing some adjustment in the tension or have to be picked instead of being raked, but it can be done. No matter how minute the clearances are, there are clearances, or the parts would not go together and this is what makes picking and raking possible. You may come across a cylinder that you cannot pick or rake in a reasonable length of time. Even the expert runs into these same problems. Do not become discouraged. Most locks can be picked or raked in a short time. Do not waste hours working on an extremely difficult cylinder. You will soon be able to determine just how much time to spend on picking or raking a cylinder before resorting to other methods such as drilling.

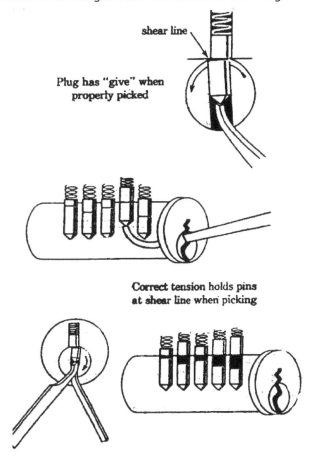

shear line

Plug has "give" when properly picked

Correct tension holds pins at shear line when picking

You will probably develop your own handling of the tools and personal grip. Tools must be comfortable in your hands. When you first start use a medium weight and length tension tool. After you have been raking for awhile you may then prefer to try a light weight or rigid tension tool. You will soon find which tool is best for you.

When working with padlocks, as you apply tension to the plug, you will be working against the direct pressure of a spring which is used in the locking of the shackle and returning of the plug to the locked position. This lock will require more tension. Use one of the rigid tools. Also if you push inward on the shackle of the padlock, it will relieve some of the spring pressure on the plug. If you have raked the plug in the wrong direction and the lock will not open, this is no problem. If the lock was very easy to pick, just apply tension in the other direction and re-rake it open. Now, if you have raked the lock in the wrong direction and opened it with difficulty, hold it right there. Do not lock it and re-rake it. There is a tool just for this purpose. It is a pair of coiled springs with handles, one coiled to the left and one coiled to the right. The first procedure is to determine if the left or right hand of the tool is required. This is done by facing the lock and looking at the handle pointing towards you and the flange inserted into the top of the keyway. If the plug is turned to the right, your handle will be to the right of the cylinder. If the plug is turned to the left, your handle will be to the left of the cylinder.

If the plug is turned to the left, and this is in the wrong direction for opening the lock, you may have to get the plug to go to the right. Carefully remove the pick and tension tool. Insert a small screwdriver into the lower portion of the keyway on the raked lock. Keep the tension on the plug with the screwdriver. Do not allow the plug to slip into the original locked position. Place the centred flange end of the coil into the upper section of the keyway. Grasping the small handle of the coil, strongly wind the flip-it toward the direction into which the plug is to be turned. This will be to the right.

Keep the plug in position firmly with the screwdriver. With a quick yank, pull back the screwdriver. The tension of the flip-it will snap the plug over to the opposite direction quickly enough to prevent the pins from falling back into their locked positions. After practicing this procedure, you will find the tool is quite easy to use. Occasionally you will come across a cylinder that rakes easier in one direction than the other and if you have to rake it into the unlocking direction, you will find this tool quite useful. When picking a cylinder you will be lifting one pin at a time. For this use a hook-type pick. Apply tension in the same manner as when raking. Insert the pick all the way into the keyway and raise up the last pin until the breaking point will bind at the shear line. Proceed to the next pin until you work your way out of the keyway. Keep tension on the plug during the entire process. After each pin is picked you will feel the bottom pin become free of the downward spring pressure. This does not mean it is picked. You may have been applying too much tension to the plug and caused the top pin to bind below the shear line.

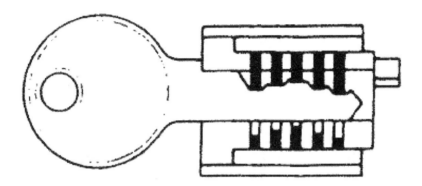

When it is properly picked you should feel a very slight give in the turning of the plug. That slight turning of the plug will become greater with each pin you pick until it turns fully when you have picked all the pins. The tension must not be too heavy. You must develop that touch with your tension hand. To practice picking do so as in the raking method. Set up a cylinder with two pins, pick it a few times, then set it up with three pins, and so on until you are able to pick a six or even a seven pin cylinder.

The raking method is all that is required for opening these locks. The raking is performed in the same manner as that of raking a pin-tumbler cylinder and they rake open quite easily. The tension is also used in the same manner as that of the pin-tumbler cylinders. The only variation would be in the double-sided disc or wafer cylinder.

The double-sided cylinder usually requires a different tension tool. Double-sided locks can be raked in two different ways:-

    1.   With the use of a standard raking tool, apply tension in the same way as you with

all other picking and raking. If you are using a double-sided tension tool it will fit in the top and bottom of the keyway.

2. After you have applied tension begin to rake the upper discs or wafers as you would have in the single-sided lock. When you feel a slight give in the tension of the plug switch raking to the bottom, but do not let up on the tension. Rake the lower discs or wafers in the same way as you did the upper ones, but use a downward pressure when pulling the rake out. It would be like raking a pin tumbler cylinder that was installed upside-down.

3. Use double-sided picks. These tools are very effective in opening most disc-tumbler double-sided locks. No tension tool is required if there is no spring tension on the plug. Spring tension will be found only in padlocks or shunt switches. Insert tip of picking tool all the way into the keyway. Rock the tool rapidly up and down while pushing slowly and gently inward. If the pick binds part way in remove tool and reverse to pick prongs on other side and try again. Do not force. When tool is inserted all the way picking action is begun by a moderately rapid up and down rocking motion together with a twisting toward the unlocking direction.

Each of the tools should be tried out using this action. Explore all possibilities until a lock opening is made. After removing the pick ensure the fork ends are straight. Re-bend to a parallel position if it becomes necessary. Most double-sided cylinders unlock to the right, clockwise. When picking the double-sided padlocks or shunt switches a heavier tension is required. The tension tool is usually required for these types of locks. Press down on the shackle while raking. This will help to relieve some of the turning tension. Both methods work very well after a little practice. You should soon become proficient in the raking and picking of pin tumblers, single and double-sided disc and wafer locks.

Lock picking can be an effective method for opening locks if certain conditions present themselves. The pins must be free and the cylinder, in general, operational. Do not let ego get in the way when determining the best method for a given situation. There is nothing wrong with drilling a lock to get it open, if that is the most time efficient and cost effective method for your customer. In most cases labour can be more expensive than the product. It is not logical to stand outside picking a lock for an hour, even if you finally do open it, when you can drill and replace most common cylinders in a few moments.

Picking is a skill that will only improve with practice, experience and dedication.

The rewards in many situations will be incalculable in terms of time and money.

LOCK TOOLS, MATERIALS, PICKS AND WRENCHES

Picks come in several shapes and sizes. The handle and tang of a pick are the same for all picks. The handle must be comfortable and the tang must be thin enough to avoid bumping pins unnecessarily. If the tang is too thin, then it will act like a spring and you will loose the feel of the tip interacting with the pins. The shape of the tip determines how easily the pick passes over the pins and what kind of feedback you get from each pin.

The design of a tip is a compromise between ease of insertion, ease of withdrawal and feel of the interaction. The half diamond tip with shallow angles is easy to insert and remove, so you can apply pressure when the pick is moving in either direction. It can quickly pick a lock that has little variation in the lengths of the key pins. If the lock requires a key that has a deep cut between two shallow cuts, the pick may not be able to push the middle pin down far enough. The half diamond pick with steep angles could deal with such a lock. In general steep angles give you better feedback about the pins. Unfortunately, the steep angles make it harder to move the pick in the lock. A tip that has a shallow front angle and a steep back angle works well for Yale locks.

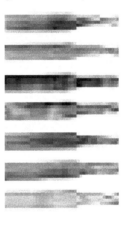

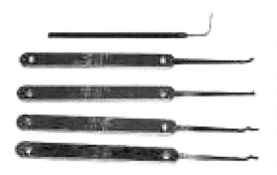

The half round tip works well in disk tumbler locks. The full diamond and full round tips are useful for locks that have pins at the top and bottom of the keyway. The rake tip is designed for picking pins one by one. It can also be used to rake over the pins, but the pressure can only be applied as the pick is withdrawn. The rake tip allows you to carefully feel each pin and apply varying amounts of pressure. Some rake tips are flat or dented on the top to makes it easier to align the pick on the pin. The primary benefit of picking pins one at a time is that you avoid scratching the pins. Scrubbing scratches the tips of the pins and the keyway and spreads metal dust throughout the lock. If you want to avoid leaving traces avoid scrubbing. The snake tip can be used for scrubbing or picking. When scrubbing the multiple bumps generate more action than a regular pick. The snake tip is particularly good at opening five pin household locks. When a snake tip is used for picking it can set two or three pins at once. Basically, the snake pick acts like a segment of a key which can be adjusted by lifting and lowering the tip, by tilting it back and forth, and by using either to top or bottom of the tip. Use moderate to heavy torque with a snake pick to allow several pins to bind at the same time. This style of picking is faster than using a rake and it leaves little evidence.

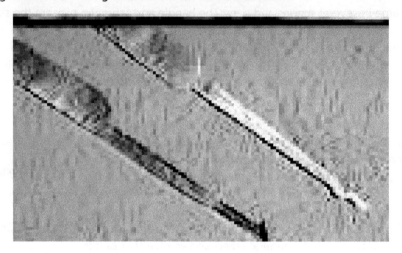

The spring steel bristles used on street cleaners make excellent tools for lock picking. The bristles have the right thickness and width and are easy to grind into the desired shape. The resulting tools are springy and strong.

The first step in making tools is to sand off any rust on the bristles. Course grit sand paper works well as does a steel wool cleaning pad (not copper wool). If the edges or tip of the bristle are worn down use a file to make them square.

A torque wrench has a head and a handle. The head is usually 1/2 to 3/4 of an inch long and the handle varies from 2 to 4 inches long. The head and the handle are separated by a bend that is about 80 degrees. The head

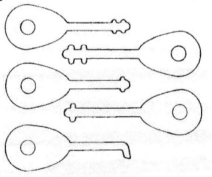

WARDED PICKS

must be long enough to reach over any protrusions, such as a grip-proof collar, and firmly engage the plug. A long handle allows delicate control over the torque. If it is too long it will bump against the doorframe. The handle, head and bend angle can be made quite small if you want to make tools that are easy to conceal (in a pen, flashlight or belt buckle). Some torque wrenches have a 90 degree twist in the handle. The twist makes it easy to control the torque by controlling how far the handle has been deflected from its rest position. The handle acts as a spring which sets the torque. The disadvantage of this method of setting the torque is that you get less feedback about the rotation of the plug. To pick difficult locks you will need to learn how to apply a steady torque via a stiff handled torque wrench.

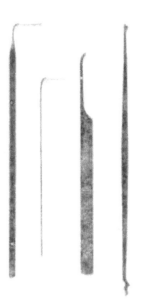

The width of the head of a torque wrench determines how well it will fit the keyway. Locks with narrow keyways, such as desk locks, need torque wrenches with narrow heads. Before bending the bristle file the head to the desired width. A general purpose wrench can be made by narrowing the tip, about 1/4 inch, of the head. The tip fits small keyways while the rest of the head is wide enough to grab a normal keyway.

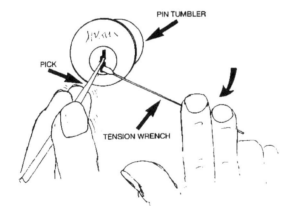

The hard part of making a torque wrench is bending the bristle without cracking it. To make the 90 degree handle twist, clamp the head of the bristle, about one inch, in a vice. Use pliers to grasp the bristle about 3/8 of an inch above the vice. You can use another pair of pliers instead of a vice. Apply a 45 degree twist. Try to keep the axis of the twist lined up with the axis of the bristle. Now move the pliers back another 3/8 inch and apply the remaining 45 degrees. You will need to twist the bristle more than 90 degrees in order to set a permanent 90 degree twist.

To make the 80 degree head bend, lift the bristle out of the vice by about 1/4 inch, so 3/4 inch is still in the vice. Place the shank of a screw driver against the bristle and bend the spring steel around it about 90 degrees. This should set a permanent 80 degree bend in the metal. Try to keep the axis of the bend perpendicular to the handle. The screwdriver shank ensures that the radius of curvature will not be too small. Any rounded object will work (a drill bit, needle nose pliers or a pen cap). If you have trouble with this method, try grasping the bristle with two pliers separated by about 1/2 inch and bend. This method produces a gentle curve that will not break the bristle.

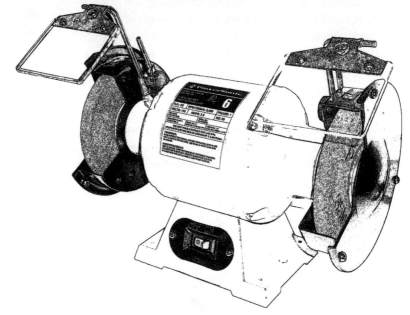

A grinding wheel will speed up the job of making a pick. It takes a bit of practice to learn how make smooth cuts with a grinding wheel, but it takes less time to practice and make two or three picks than it does to hand file a single pick. The first step is to cut the front angle of the pick. Use the front of the wheel to do this. Hold the bristle at 45 degrees to the wheel and move the bristle side to side as you grind away the metal. Grind slowly to avoid overheating the metal, which makes it brittle. If the metal changes to dark blue you have overheated it and you should grind away the coloured portion. Next, cut the back angle of the tip using the corner of the wheel. Usually one corner is sharper than the other and you should use that one. Hold the pick at the desired angle and slowly push it into the corner of the wheel. The side of the stone should cut the back angle. Ensure the tip of the pick is supported. If the grinding wheel stage is not close enough to the wheel to support the tip use needle nose pliers to hold the tip. The cut should pass though about two thirds of the width of the bristle. If the tip came out well, continue. Otherwise break it off and try again. You can break the bristle by clamping it into a vice and bending it sharply.

The corner of the wheel is also used to grind the tang of the pick. Put a scratch mark to indicate how far back the tang should go. The tang should be long enough to allow the tip to pass over the back pin of a seven pin lock. Cut the tang by making several smooth passes over the corner. Each pass starts at the tip and moves to the scratch mark. Try to remove less than a 1/16th of an inch of metal with each pass. Use two fingers to hold the bristle on the stage at the proper angle while your other hand pushes the handle of the pick to move the tang along the corner.

Use a hand file to finish the pick. It should feel smooth if you run a finger nail over it. Any roughness will add noise to the feedback you want to get from the lock.

The outer sheath of phone cable can be used as a handle for the pick. Remove three or four of the wires from a length of cable and push it over the pick. If the sheath will not stay in place put some epoxy on the handle before pushing the sheath over it.

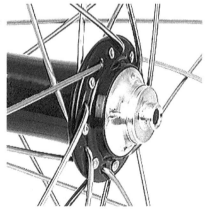

An alternative to making tools out of street cleaner bristles is to make them out of nails and bicycle spokes. These materials are easily available. When they are heat treated they will be stronger than tools made from bristles.

A strong torque wrench can be constructed from a nail (about .1 inch diameter). Heat up the point with a propane torch until it glows red, slowly remove it from the flame and let it air cool. This softens it. The burner of a gas stove can be used instead of a torch. Grind it down into the shape of a skinny screwdriver blade and bend it to about 80 degrees. The bend should be less than a right angle because some lock faces are recessed behind a plate (called an escutcheon) and you want the head of the wrench to be able to reach about half an inch into the plug. Temper (harden) the torque wrench by heating to bright orange and dunking it into ice water. You will wind up with a virtually indestructible bent screwdriver that will last for years with extreme use.

Bicycle spokes make excellent picks. Bend one to the shape you want and file the sides of the business end flat so it is strong in the vertical and flexible in the horizontal direction. For smaller picks, which you need for very small keyways, find any large diameter spring and unbend it.

Slim Jim

Slim jims are used for unlocking car doors when there is no key present. Many people are very thankful for the existence of slim jims when their keys are beyond reach.

[[IMAGE SLIM JIM]]

One must possess extensive expertise to use one of these devices. Improper use of a slim jim can wreak havoc on the locking mechanism in a car door. Either using an

incorrect model or using a correct model incorrectly can cause lasting damage to car locks. If you are a beginner looking to use these products, make sure you have the proper training and practice under your belt before using one of these tools on a car.

Tubular Lock Picks

Tubular lock picks are designed to pick complex tubular locks. Traditional methods of lock picking rarely work effectively on these kinds of locks. Tubular locks can present a challenge.

A standard tubular lock pick will only contain seven or eight pins. The name comes from their cylindrical shape, which matches that of the locks that they pick. Tubular lock picks have been used since the 1930s to pick locks on most coin-operated vending machines, laundry machines, cigarette machines, jukeboxes and pinball machines.

Using Tubular Lock Picks

Tubular pin tumbler locks are generally considered to be safer and more pick-resistant than standard pin-tumbler locks. Tubular locks are found on many vending machines such as coin-operated washers and dryers, bicycle locks, and are even used in many retail stores to lock jewellery showcases. Tubular lock picks are made to compromise the 7 or 8 pins present in a given tubular lock. Tubular lock picks are specialised tools and require more skill to use than standard lock picks.

Tubular picks are inserted into the lock and turned clockwise with light to medium tension. As the tool is gently pushed into the lock, each of the picks is slowly forced down until they stop, binding the driver pins behind the shear line of the lock. When the final pick is pushed down, you have aligned the pin segments with the shear line allowing the lock to open.

Electric Lock Picks

Electric lock picks are made for professionals who need to open a large number of the same type of lock in quick succession. Property managers, security guards and storage facility managers and owners find electric lock picks and pick guns to be invaluable in their every day work. Most recreational lock pickers engage in this activity due to its challenging nature. Electric picks essentially take the challenge out of picking a lock, but they also make the job much easier for an unskilled user.

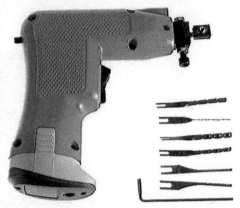

For professionals, however, electric lock picks can be very valuable tools. Electric picks work in a rather unscientific way. They vibrate when inserted into a lock, and the motion often causes pins to fall into place in such a way that allows the lock to be released. Such a method of lock picking may not be very challenging, but it can be immensely rewarding for the busy professional locksmith. Make sure your electric pick is made of the finest steel and has a good warranty. It is going to get a lot of use.

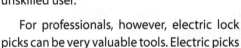

The Snap Pick and lock Pick Gun

Another means of opening locks without the key is by using a snap pick, lock pick gun or vibrating tool. These devices all exploit Newton's law that says that for every action there is an equal and opposite reaction. If a ball all the way on the left or right side is lifted up and let loose to collide with the row of suspended balls, this ball will transfer all its energy to the next ball and so forth, until the ball on the other end moves to swing away from the other balls. When it swings back, the process is reversed and the original ball swings up.

This principle can be used to open locks. If impulse energy is transferred to the first pin, it will tend to stay in place and the second pin tends to move away from the first one, until the spring stops it and pushes it back to touch the first pin.

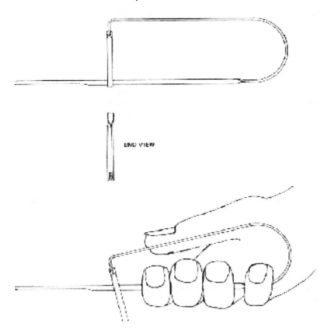

END VIEW

A lock pick gun will, when the trigger is pulled, tension a spring and then when the trigger is pulled all the way use the force of that spring to snap the needle up for a short distance. By positioning this needle into the lock, just touching the pins and then pulling the trigger, one tries to hit all the pins simultaneously. By then making the lock turn in the split-second before all the upper pins are pushed back by the springs in the lock, one can open the lock. The amount of turning force and the timing with which to apply it require some training.

Vibrating picks use the same principle except many times a second, requiring less training on the part of the operator. A snapper pick is the simpler version of a pick gun. The lock industry has created locks that are more resistant to this technique. More resistant locks have narrower keyways, preventing tools from being inserted in the first place. This makes it harder to transfer the impulse energy to the pins. More resistant locks also have smaller tolerances, creating less space for the pins to bounce around.

Exercises how to Improve Skills in using Lock Picks and Tools.

Applying a Fixed Pressure with a Pick

This exercise will help you learn the skill of applying a fixed pressure with the pick, independent of how the pick moves up and down in the lock. Basically you need to learn how to let the pick bounce up and down according to the resistance offered by each pin. How you hold the pick makes a difference on how easy it is to apply a fixed pressure. You must hold it in such a way that the pressure comes from your fingers or your wrist. Your elbow and shoulder do not have the dexterity required to pick locks. While you are scrubbing a lock, notice which of your joints are fixed and which are allowed to move. The moving joints are providing the pressure.

One way to hold a pick is to use two fingers to provide a pivot point, while another finger levers the pick to provide the pressure. Which fingers you use is a matter of personal choice. Another way to hold the pick is like holding a pencil. With this method your wrist provides the pressure. If your wrist is providing the pressure, your shoulder and elbow should provide the force to move the pick in and out of the lock. Do not use your wrist to both move the pick and apply pressure.

To get used to the feel of the pick bouncing up and down in the keyway try scrubbing over the pins of an open lock. The pins cannot be pushed down so the pick must adjust to the heights of the pins. Try to feel the pins rattle as the pick moves over them. If you move the pick quickly, you can hear the rattle. This same rattling feel will help you recognise when a pin is set correctly. If a pin appears to be set but it does not rattle, then it is false set. False set pins can be fixed by pushing them down farther, or by releasing torque and letting them pop back to their initial position.

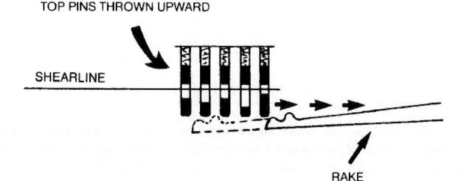

TOP PINS THROWN UPWARD

SHEARLINE

RAKE

Range of Pressures to Apply with a Pick

This exercise will teach you the range of pressures you will need to apply with a pick. When you are starting, just apply pressure when you are drawing the pick out of the lock. Once you have mastered that try applying pressure when the pick is moving inward.

With the flat side of your pick, push down on the first pin of a lock. Do not apply any torque to the lock. The amount of pressure you are applying should be just enough to overcome the spring force. This force gives you an idea of minimum pressure you will apply with a pick. The spring force increases as you push the pin down. See if you can feel this increase.

Now see how it feels to push down the other pins as you pull the pick out of the lock. Start out with both the pick and torque wrench in the lock, but do not apply any torque. As you draw the pick out of the lock, apply enough pressure to push each pin all the way down.

The pins should spring back as the pick goes past them. Notice the sound that the pins make as they spring back and the popping feel as a pick goes past each pin. Notice the springy feel as the pick pushes down on each new pin. To help you focus on these sensations, try counting the number of pins in the lock. To get an idea of the maximum pressure, use the flat side of your pick to push down all the pins in the lock. Sometimes you will need to apply this much pressure to a single pin.

Range of Torque to apply to a Lock

This exercise will teach you the range of torque you will need to apply to a lock. It demonstrates the interaction between torque and pressure. The minimum torque you will use is just enough to overcome the friction of rotating the plug in the hull. Use your torque wrench to rotate the plug until it stops. Notice how much torque is needed to move the plug before the pins bind. This force can be quite high for locks that have been left out in the rain. The minimum torque for padlocks includes the force of a spring that is attached between the plug and the shackle bolt.

To get a feel for the maximum value of torque, use the flat side of the pick to push all the pins down. Try applying enough torque to make the pins stay down after the pick is removed. If your torque wrench has a twist in it you may not be able to hold down more than a few pins.

If you use too much torque and too much pressure you can get into a situation like the one you just created. The key pins are pushed too far into the hull and the torque is sufficient to hold them there. The range of picking torque can be found by gradually increasing the torque while scrub¬bing the pins with the pick. Some of the pins will become harder to push down. Gradually increase the torque until some of the pin set. These pins will lose their springiness. Keep¬ing the torque fixed, use the pick to scrub the pins a few times to see if other pins will set.

The most common mistake of beginners is to use too much torque. Use this exercise to find the minimum torque required to pick the lock.

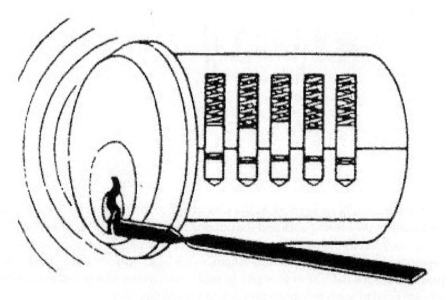

Identify which Pins are Set

While you are picking a lock try to identify which pins are set. You can tell a pin is set because it will have a slight give. The pin can be pushed down a short distance with a light pressure, but it becomes hard to move after that distance. When you remove the light pressure the pin springs back up slightly. Set pins also rattle if you flick them with the pick. Try listening for that sound. Run the pick over the pins and try to decide whether the pins are in the front or back of the lock, or both. Try to identify exactly which pins are set. Pin number one is the front most pin. The most important skill of lock picking is the ability to recognise correctly set pins

Try repeating this exercise with the plug turning in the other direction. If the front pins set when the plug is turned one way, the back pins will set when the plug is turned the other way. One way to verify how many pins are set is to release the torque and count the clicks as the pins snap back to their initial position. Try to notice the difference in sound between the snap of a single pin and the snap of two pins at once. A pin that has been false set will also make a snapping sound.

Try this exercise with different amounts of torque and pressure. Notice that a larger torque requires a larger pressure to make pins set correctly. If the pressure is too high the pins will be jammed into the hull and stay there.

Visual Skills

This exercise requires a lock that you find easy to pick. It will help you refine the visual skills you need to master lock picking. Pick the lock and try to remember how the process felt. Rehearse in your mind how everything feels when the lock is picked properly.

Basically, you want to create a memory that records the process of picking the lock. Remember the motion of your muscles as they applied the correct pressure and torque and feel the resistance encountered by the pick. Now pick the lock again trying to match your actions of the memory. By repeating this exercise, you are learning how to formulate detailed commands for your muscles and how to interpret feedback from your senses. The mental rehearsal teaches you how to build a visual understanding of the lock and how to recognise the major steps of picking it.

ADVANCED AND AUTOMATIC LOCK GUNS

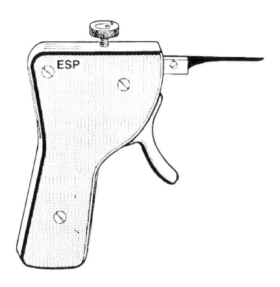

Typical of automatic lock tools is the Lockaid pick gun, invented in 1935. It was closely controlled by the authorities and each one was registered and sold only to locksmiths and police officials. The guns resembled toy cap guns and were riveted closed. Both hands are required to open locks with this instrument. A small black tension wrench is supplied with each gun. Do not attempt to open locks without this wrench. Pin-tumbler-type locks have a cylinder core or round shaft that rotates to the left (anticlockwise) or right (clockwise) after inserting a key in the keyway. In the centre of this cylinder core or shaft is the keyhole or keyway, which is an oblong slit in a vertical position. When opening this type of lock first insert the key, turn the key to the right or left and the lock is opened. By inserting the key and not turning it, the lock will never open.

It is the same with the Lock Gun. After inserting the gun needle in the keyway and pulling the trigger, the lock will not open unless the keyway turns to the right or left.

Since the needle would break if you tried to turn the keyway, it is necessary to use the small tension wrench at this point.

When you trigger the gun the upward blows of the needle knocks the tumblers in place so the shaft or cylinder core will turn freely and open the latch. The turning of the shaft or core when using the lock gun is done by turning the tension wrench.

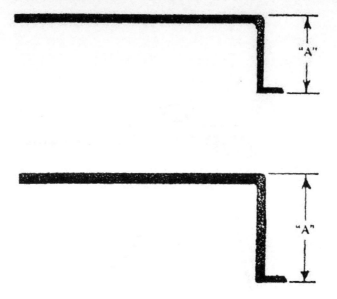

Before attempting to open a lock with the gun determine whether the keyway turns to the right or left. A good many tumbler locks turn either way to open. Others turn only to the right, while others to the left. It is safe to assume that the keyway should be turned to the right. If this does not work, use the tension wrench so the keyway turns to the left.

Step-by-Step Procedure

1. A knurled wheel at the back of the gun takes care of the tension adjustment of the blow. The wheel is turned toward you for greater tension, and away from you for less tension. Before starting to work on a lock, turn the wheel away from you as far as it will go. Then turn the wheel toward you five turns, which will produce a little tension.

2. First insert one curved end of the tension wrench at the bottom of the keyway. Arrange one curved end

of the wrench in the keyway so the other end will hang slightly to the left. In this position the wrench will act as a lever, permitting you to turn the core after the tumblers are knocked into place.

3. Insert the gun needle in the keyway directly above the end of the wrench. Keep the needle as low as possible in the keyway under the tumbler pins. When inserting the gun needle, keep it in a straight line. Do not push the gun needle into the keyway too far. Look at one of your keys, the notches should give you an idea of how far to insert the needle. Inserting the needle too far will catch the inside end of the lock and fail to strike the pins.

4. While holding the gun free in the lock with one hand maintain a slight pressure on the wrench with a finger of your left hand. Only a slight touch on the wrench is necessary. Do not exert heavy pressure on the wrench.

5. Pull the trigger while holding the gun needle in a straight line and pull slowly. After each shot, apply a slight finger pressure to the wrench with your left finger. Then release the finger pressure on the wrench.

6. When the gun needle knocks the pins into position inside the lock you will feel a slight give on the wrench. This means the gun has done its work and you should immediately stop triggering. Turn the wrench slowly to the right as you would a key and the shaft will turn and open the lock.

7. Using the wrench as a lever push your finger to the right. The keyway will turn with the wrench and the lock will open.

8. It is going to be slightly awkward to keep the curved end of the wrench in place in the bottom of the keyway without using too much pressure with your finger or fingers. However, after a few trials, this awkwardness will disappear.

9. Try both curved ends in the bottom of the keyway. One end is slightly narrower than the other.

10. The wrench can also be used at the top of the keyway but this is more awkward.

11. If the lock does not open after eight or nine shots release the tension on the wrench, allowing the pins to drop back into place, and start again. If the lock still does not open keep turning the wheel of the gun toward you for more tension. Keep a record of the number of turns it takes to successfully open a lock as this will serve as a guide for similar locks.

12. Some locks have narrow corrugated keyways which mean that the gun blade or needle will have to be filed thinner. Never file the top edge of the blade or needle. Only file the sides or bottom.

13. An additional offset gun needle is furnished with each gun. This is used only on locks that are directly on top of the doorknob, in which case the needle cannot be held in the keyway in a straight line.

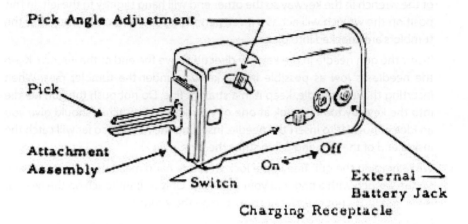

Pick Angle Adjustment

Pick

Attachment
Assembly

Switch

On → Off

External
Battery Jack

Charging Receptacle

GENERAL VIEW OF DEVICE

Electric variants of the automatic pick gun soon followed. Electric lock pick guns are for use in opening pin tumbler locks. You should never purchase an electric pick in the hopes that it will replace an entire professional pick set. Electric lock picks are expensive and are not of very much value unless you have a succession of the same type of pin tumbler locks to open quickly. There are law enforcement professionals, property managers and storage facility managers that choose to use an electric pick gun because they are easier to use than a standard lock pick set. You insert these long pieces of metal (needles) into the lock, in the same way as you would insert a regular lock pick. As the metal pieces vibrate, they push the pins up. This works something like raking a lock, but much faster. Sometimes electric picks will open the lock in just a few of seconds and sometimes they will not work

at all. Most hobbyists avoid electric lock picks because they take the challenge out of the process. To them, if there is no challenge in defeating a lock it s not truly lock picking. To other security professionals, the electric lock pick gun is invaluable.

A tubular pin tumbler lock, also known as Ace lock, axial pin tumbler lock or radial lock, is a variety of pin tumbler lock in which 6-8 pins are arranged in a circular pattern. The corresponding key is tubular or cylindrical in shape.

Tubular locks are commonly seen on bicycle locks, computer locks, escalators and a variety of coin-operated devices such as vending machines, coin-operated washing machines and slot machines.

Tubular pin tumbler locks are generally considered to be safer and more resistant to picking than standard locks, though there are several ways to open them without a key. Even though the pins are exposed, making them superficially easier to pick, they are designed such that after all pins are manipulated to their shear line, once the plug is rotated 1⁄6 to 1⁄8 around, the pins will fall into the next pin's hole, requiring re-picking to continue. As such, picking the lock without using a device to hold its pins in place once they reach their shear line requires one complete pick per pin.

Such locks can be picked by a special tubular lock pick with a minimum of effort in very little time. It is also possible to defeat them by drilling with a hole saw drill bit. Standard tubular lock drill bit sizes are 0.375 in (9.5 mm) diameter and 0.394 in (10.0 mm) diameter. To prevent drilling many tubular locks have a middle pin made of hardened steel or contain a ball bearing in the middle pin.

A tubular lock pick is a specialised lock picking tool used for opening a tubular pin tumbler lock. Tubular lock picks are all very similar in design and come in sizes to fit all major tubular locks, including 6, 7 and 8 pin locks.

The tool is simply inserted into the lock and turned clockwise with medium tension. As the tool is pushed into the lock each of the pins is slowly forced down until they stop. This binding the driver pins behind the shear line of the lock. When the final pick is pushed down the shear plane is clear and the lock opens. This can usually be accomplished in a matter of seconds.

Combination Locks

Combination locks work on a series of flat round disks that have notches and pegs (one of each, one set per disk) along their circumference. Notches are referred to as gates. The first tumbler determines the last digit of the combination and is actually attached to the dial directly. As the dial is turned the peg of the first tumbler catches on the middle tumbler's peg, dragging it along. As the dial is turned further the middle tumbler latches on to the peg of the last tumbler, all three turning together. Turning all the tumblers is known as clearing the lock. This must be done before attempting to operate the lock. For the lock to open, the gate on each disk must align up with the pawl (breaking arm) of the bolt. Dialling the first digit of the combination aligns the last tumbler's gate to the pawl. Before dialling the second digit the dial must be turned one complete turn in the opposite direction (assuming a three tumbler lock, twice for a four digit one). Rotating in the original direction to the last digit will align the first tumbler's gate, and the lock can open. Modern safe combination locks are impossible to crack (literally). Many innovations have given high quality locks this degree of security.

Burglars learned to feel the gates and pegs rotate about the lock, allowing them to manipulate the tumblers into their proper position. To combat this, a serrated front tumbler was designed to create shallow false gates. The false gates are difficult to distinguish from the actual gates. To combat this problem, safe crackers would hook up a high speed drill to the dial. This would wear the tumblers' edges smooth, eliminating the bothersome shallow gates.

Despite their security, cheap combination locks are far from foolproof. The most common and difficult to open of these small disk tumbler locks are the Master combination padlocks, and they are quite popular. With practice, they CAN be opened. The newer the lock is, though, the more difficult it will be to open at first. If the lock has had a lot of use, such as that on a locker-room door, where the shackle gets pulled down and encounters the tumblers while the combination is being dialled, the serrated front tumblers will become smoothed down. This allows easier sensing of the tumblers. Until you have become good at opening these locks practice extensively on an old one.

Here is how. First, clear the tumblers by engaging all of them. This is done by turning the dial clockwise. Sometimes these locks open more easily starting in the opposite direction three to four times. Now bring your ear close to the lock and gently press the bottom back edge to the bony area just forward of your ear canal opening, so that vibrations can be heard and felt. Slowly turn the dial in the opposite direction. As you turn you will hear a very light click as each tumbler is picked up by the previous tumbler. This is the sound of the pickup pegs on each disk as they engage each other. Clear the tumblers again in a clockwise manner and proceed to step two. After you have cleared the tumblers, apply an upward pressure on the shackle of the padlock. Keeping your ear on the lock, try to hear the tumblers as they rub across the pawl. Keep the dial rotating in a clockwise direction. You will hear two types of clicks, each with a subtle difference in pitch. The shallow, higher pitched clicks are the sound of the false gates on the first disk tumbler. The real gates sound hollow and empty, almost non existent. When you feel a greater than normal relief in the shackle once every full turn, this is the gate of the first tumbler (last number dialled). This tumbler is connected directly to the dial. Ignore that sound for now. When you have aligned the other two tumblers, the last tumbler's sound will be drowned out by the sound of the shackle popping open. While continuing in a clockwise direction with the dial, listen carefully for the slight hollow sound of either one of the first two tumblers.

Note on the dial face where these sounds are by either memorising them or writing them down. Make certain that you do not take note of the driving tumbler (last number dialled). If you hear and feel only one hollow click (sounds like dumpf), chances are that the first number could be the same as the last one. You should have two numbers now. Let us say one of them is 12 and the other is 26. Clear the tumblers again just to be safe and stop at the number 12. Go anticlockwise one complete turn from 12. Continue until there is another dumpf sound. After the complete turn pass 12. If you feel and hear a louder than normal sound of a tumbler rubbing on the pawl, the first tumbler is properly aligned and the second tumbler is taking the brunt of the force from the shackle, you are on the right track. When the second tumbler has aligned in this case, you will feel a definite resistance with the last turn of the dial going clockwise. The final turn will automatically open the shackle of the lock. If none of these symptoms are evident, try starting with the number of the combination, 26, in the same way.

Do not give up if the lock still does not open. Try searching for a different first number. Try for up to forty minutes. If you play with it long enough, it will eventually open. The more practice you have the quicker you will be able to open these padlocks in the future. Using a stethoscope to increase audibility of the clicks is not out of the question when working on disk tumbler locks, though usually not needed for padlocks. A miniature wide audio range electronic stethoscope with a magnetic base for coupling a piezoelectric-type microphone is ideal for getting to know the tumblers better. Most people are aware of the complex equipment and methods for picking locks, but not many people are aware of very simple methods such as card-picking.

The first trick is to be able to determine what sort of door it will and will not work on. The type of door of you are looking for is one that will move back and forth slightly when you push and pull on the door handle, and one that does not have an over-hanging door frame. Take a laminated card, credit cards are fine, and push it in between the door and door frame as you lean against the door. You should push the card in adjacent to the door handle, as hard against the inner bolt as you can. When you slowly lean back off the door the bolt should slide up onto your card and the door will come open.

Another skill essential for the locksmith is key impressioning. While not an advanced technique as such, it is vital for the locksmith to practice this skill alongside lock picks and electric pick guns.

You can start out with any lock but a good suggestion is an average sized four-pin Master padlock. They are easy to impression and blanks can be easily obtained at a hardware store. You should get more than one blank for practicing. Five is probably a reasonable number.

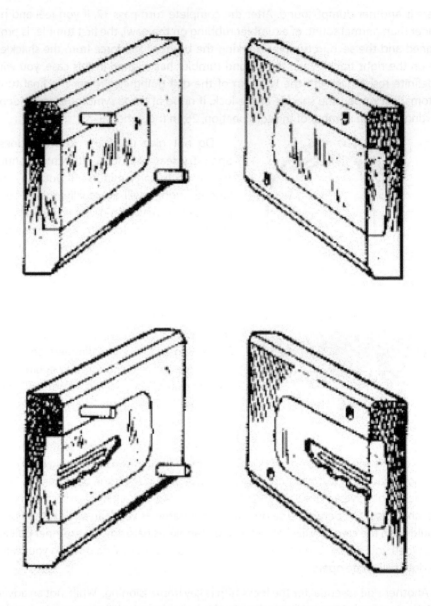

Below are some depths and spacing data for the Master padlock. The depths are measured from the bottom of the blade of the key, up to the bottom of the cut where a pin will rest.

Cut number	Depth	Cut number	Depth
0	.280″	4	.220″
1	.265″	5	.205″
2	.250″	6	.190″
3	.235″	7	.175″

The distance from the shoulder of the key to the first pin is .185″ and the spacing between pins is .125″. You do not need these last two numbers, but they may be helpful references as you are first learning to recognise what the marks look like. Another good approach to using a practice lock is similar to that sometimes recommended for learning picking. Get a lock cylinder and remove all the pin stacks but one. After you have impressioned the one pin lock, add another pin stack and try again. Continue adding pin stacks until you can impression the whole lock.

You will need a stock of files for making and adjusting keys. A six inch, number 4 Swiss-cut round or pippin files are normally used for impressioning. The files are called six inch but are actually about eight inches long including the tang. Both types of files taper down to a smaller cross-sectional size towards the tip. The round file is usually used in conjunction with a small flat or triangular file which is used to shape the flat sides of the cuts on a key. The pippin file has a teardrop cross section, rounded on one side and with two flat surfaces meeting at a knife edge on the other side. The flat surfaces are used like the flat file above to shape the sides of cuts.

The particular number 4 Swiss-cut pattern file is used for impressioning work because it leaves a very fine, slightly dull and slightly corrugated surface on the blank. This permits visible marks to be made by the pins rubbing on the blank with very little pressure. A few locksmiths use a number 2 Swiss cut pattern because it cuts faster. It is a good idea to get a handle for the file, as it permits better and more comfortable control of the file. A file card is a special brush made to clean the teeth of a file. The soft brass of the key blanks tends to clog up the teeth on an impressioning file which affects the quality of the fine surface you are trying to produce on the blank.

Files cut only on the forward stroke. So push the file slowly and evenly forward with gentle cutting pressure and draw back the file without any cutting pressure. Particularly when impressioning do not apply pressure when drawing back the file, as it tends to polish the surface of the blank (a dull surface is needed when impressioning). Hold the file with an extended index finger pushing down on the top edge of the file to control cutting pressure. Light cutting pressure will produce the finest finish for producing visible marks. Use heavier pressure to remove material rapidly, followed by lighter strokes to finish the surface for marking.

Soft brass blanks are the best for impressioning. Steel blanks are much harder and aluminum blanks develop fatigue cracks easily when using hard turning tension. If you can only find bright plated brass blanks you will have to file the plating off the top of the blade with your impressioning file. Only file deep enough to remove the plating because with some locks a number 0 cut requires the full uncut height of the blade. With the plain brass blanks you need to smooth the top of the blade with your impressioning file in order to leave a surface that will show marks. Be careful not to take off too much. Some lock manufacturers use number 0 and others use number 1 to indicate the highest depth cut.

Some people like to prepare the blanks by either thinning them down in width with a flat file, or knife edging the top of the blade. In both cases the idea is that a very thin piece of metal can more easily be deformed than a thick one. In the case of thinning down the blade, it can also be wiggled around more in the keyway. When thinning a blade do not thin the area immediately adjacent to the shoulder of the blank where the blade enters the keyway. You will be applying hard turning tension on the blank later and it is important not to weaken it at the point where most of the turning stress is applied.

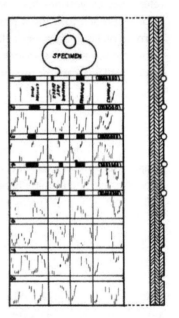

Knife edging is used more often when the pull-out method of obtaining marks is to be used. Knife edging is used to thin only the top of the blade to make the initial marks more visible. To knife edge the blade, file both sides of the top the blade at about a 45 degree angle. The idea is not to make it really sharp as a knife, just to make the edge weak enough to mark more easily on the top surface.

There are three commonly used methods for making the marks.

They are called:-

- wiggling
- tapping
- pulling

In each of the methods the blank is inserted in the keyway, then turned hard to bind the pins. Usually turning pressure is applied in the direction you want the lock to open, but you can try both directions to see which leaves better marks. It is important to make sure that the blank is evenly seated on the bottom of the keyway before applying turning pressure. If you are holding it tilted some of the pins will already be pushed up and will not leave any marks.

When impressioning you will need something to hold the blank because of the repeated hard turning tension used. The tension is harder than is used for picking, but not hard enough to break the blank. A small pair of vice-grips, no larger than the 5" size, works well. Attach the vice-grips like a handle, aligned with the long the axis of the key blade, not at a right angle like a turning wrench. There are also some commercially made handles for impressioning. There is at least one with a trigger handle to help pull out the blank uniformly each time when using the pull-out method.

Wiggling is accomplished by applying turning tension, then wiggling the blank up and down causing the top of the blank to rub against the tips of the bound lower pins.

Tapping is a variation of wiggling. The blank is inserted into the keyway then a steel rod is placed in the hole in the bow (handle) of the key to provide turning tension. A small mallet is used to tap on the bow to make the impressions. Tapping on the top of the bow pushes up the tip of the key by lever action. Tapping on the bottom of the bow pushes up the back of the key by direct action.

The pull-out method only works after you have cut down to at least a number 1 depth. Hence the popularity of knife edging the blank, then use the wiggle method to see if there are any number 0 cuts to start with. To use the pull-out method apply turning tension then pull out on the blank. Do not try this method on disk or wafer locks because the disks may bend or break. Unlike the wiggle and tapping methods the marks produced by pulling will not be exactly where the pins are, the distance away being related to how far you pull the blank out (maybe 1/16"). For this reason it is helpful to scribe lines down the side of the blank after the pin locations are found by the wiggle method, to use as a reference when filing. The advantage of the pull-out method is that it can leave more easily visible marks than the previously mentioned methods.

There is more than one way to implement the pull-out method. One technique involves attaching a C-clamp to the bow, then using the C-clamp to provide turning tension on the blank. A screwdriver is placed between the side of the bottom end of the C-clamp and the face of the lock. Then the screwdriver is twisted to pry the C-clamp, and therefore the blank, in a direction out from the face of the lock (no more than about 1/16").

An effective hybrid approach is to first put turning pressure on the blank then add pulling pressure without actually pulling the blank out enough to start making marks. The pressure is just to take up any slack between the blank and the pins and to put more tension on the pins. Using vice-grips or a commercial impressioning handle then bump or tap the blank up and down to make the marks stand out more than more than they would otherwise. Remember to file where the pins are, as with other pull-out techniques. There is an optimum amount of turning tension to apply to the blank for any particular lock. It is the rubbing action of the pins against the blank that polishes the surface of the blank to produce the little marks used for impressioning. If too little tension is used, the pins will move too easily and not mark. If too much turning tension is used the pins will jam and not mark. The pins have to be able to move a little to polish the blank's surface.

You will have better control of the impressioning action if you hold the blank and handle with your hand up near the head of the blank and the face of the lock, rather than having your hand farther away.

Wrist action, rather than action from the elbow, is more effective in moving the blank within the keyway to produce marks. The recommended action is to tilt the key up and down from the wrist with a bit of a snap, verses just lifting and lowering the blank.

The mere act of preparing the flat top of a soft brass blank with an impressioning file, inserting the blank in a lock and removing it without any wiggling or turning, will leave marks on the blank. There will be some streak marks where the pins have dragged across the specially prepared surface. Try it and you will know these marks look like, so you will not confuse them later with the useful marks.

The useful marks you get are not really depressions in the surface of the blank, except maybe when a pin is almost at the shear line. If you start seeing deep gouges, the lock is probably about to open. A mark is normally just a subtle change in the reflectivity of the surface of the blank. The impressioning file leaves a slightly dull finish and marking will slightly polish it. To see the marks turn the blank in the light. When you hold it at the right angle the marks appear as tiny shiny dots. They can be hard to see in bright light, so if working outdoors sun glasses may be helpful. Some people like to use a magnifier to see the tiny dots. Even with a magnifier you still have to turn the blank in the light just right to see the marks. With a little practice you will locate the marks very quickly.

If impressioning a dirty or weathered lock, you may find little specks of debris on the surface of the blank after marking. If there is any doubt as to what you are looking at, wipe off the top of the blank to see if you actually have a mark rather than a tiny speck of dirt.

The rule for filing marks is simple. If you see a mark, you file there. If not, you do not, except when using the pull-out method, in which case if you see a mark file where the pins are. Whatever you do, do not be tempted to guess. If you are unsure if you have a mark, do not file there. Work on making and seeing the marks first.

File only two or three strokes at a time before looking for more marks. You only have to file a cut a few thousandths of an inch too deep to pass by the shear line. As the cuts are filed deeper the sides of the cuts will start to become parallel with each other, looking something like the letter U. If you leave them that way the key will get stuck in the lock.

Use a flat file, or the flat side of your pippin file, to angle the sides the cuts at about a 45 degree angle from vertical making the sides of the cuts look more like the letter V. The bottoms of the cuts should remain rounded. It can be helpful to look at some other keys then try to duplicate the shape of the cuts.

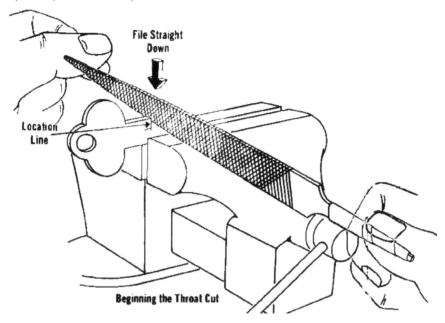

Some locks have fat pins and some have skinnier pins. There seems to be a natural tendency to use the middle part of the file, leaving fairly wide cuts. The cuts only need to have a radius a little bigger than the radius of the pin tips. For locks with skinny pins use the file more towards the tip where it is narrower. If you can see more than one mark at a time file them all at once or one at a time. Sometimes a pin will stop marking before it reaches the shear line. Do not be surprised when a pin that has stopped marking starts marking again after some of the other pins have been brought to down the shear line. Just keep filing until the pin stops marking again.

For factory keyed locks only certain standard pin depths are used. The standard pin depths are listed in depth and spacing manuals and code books available from locksmith suppliers. You can calculate the standard depths, within certain tolerances, by measuring the cut depths on other keys for the same type of lock you are working on. If you think a lock is keyed to factory depths there is no reason to look for new marks after only two or three file strokes. If you get a mark at some standard depth number then just file down the cut to the next standard depth and look for marks again.

It is helpful to have a key micrometer or dial caliper to measure the depths. A key machine can be used to speed up the impressioning process by quickly cutting down to

LOCKSMITHING, LOCK PICKING & LOCK OPENING

the next standard depth. Punch type code machines are especially useful out in the field. If you use a machine to make the cuts you will need to lightly touch up the surface of the cut with your impressioning file before looking for more marks.

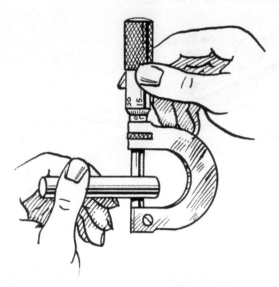

Some locks use short pins. When short pins are present you can look into the keyway and see the dividing line between the upper and lower pins. It is possible to read the short pins to determine the depth of cuts needed on the blank without any impressioning being required. To identify the short pins visually use a torch (or an otoscope, if you have one) and a straight pick. Lift up all the pins as high as they will go, then look into the keyway. Withdraw the pick slowly to drop the pins one at a time. If you see the dividing line on a pin stack, depending on its position in the keyway, you can estimate the depth of cut for that pin from your experience with other locks of the same type without doing any impressioning all. If you are familiar with the particular type of lock and do not see any dividing lines at all, you can start by filing down all the cuts to perhaps the number 1 depth or maybe the number 2 depth.

Another way to find the short pins is to use a probe. The probe is a straight pick filed to a sharp point. To use the probe lift up all the pins as high as they will go, then pull out the probe until the last pin drops. Slide the probe down the side of the pin and stop if you feel a dividing point between upper and lower pins. Note at which reference dot the probe is at, then push it all the way into the gap between the upper and lower pins. Note how much farther the probe has moved into the keyway. By measuring how far you can push the probe into the gap you can measure the size of the gap, and therefore determine the cut depth for that particular pin. Repeat the process for each pin. . The Master padlock can also be probed. Probing can also be used to assist picking. If you can tell which pins are short and which are longer before you start picking, you will have a better idea how you are going to need to manipulate the pins. Probing will leave little scratches on the side of the pins, but it does not damage the lock.

Upper spool pins are no problem because the upper pins never go below the shear line when impressioning. A few locks have lower spool pins. Using the probe you can often feel the shoulder of the spool, which feels different than a short pin because of its shape. If you find a lower spool pin file down the cut for that pin until it stops marking. Impression all the other pins normally. When only the spool in is left to be impressioned the plug will turn a little and catch on the spool pin. At that point file down the cut for the spool pin until it starts to mark again. Then continue filing it just a little bit more to bring the dividing line between the upper and lower pins down into alignment with the shear line.

Sometimes due to the stresses encountered during impressioning a blank will start to crack, usually on the blade near the shoulder where it just enters the lock. If this happens, stop. You do not want a broken-off key in the lock. You can duplicate the cracked blank on a machine or by hand, then continue impressioning with the new blank. If you do not have a key machine or a key micrometer for duplicating the cracked key, there is an old method you can use. Smoke the blade of the cracked key blank over a candle, covering it with soot. Clamp it next to a new blank using a vice, C-clamp, vice-grips, etc. Then file down the new blank until you just start to hit the soot on the old key blank. As soon as you start to scrape off the soot, stop filing. It is important not to go too deep.

If you are cracking blanks more than occasionally, you probably are using too much turning pressure. You need just enough pressure to make the marks. Turning and wiggling a blank in one direction then turning the other way and wiggling again, tends to fatigue the blank faster than working in only one direction. Watch for cracks if you are using both directions. If you accidentally make a cut a little too deep there a couple of ways you can try to save the blank. It can be peened with a small hammer or pin punch on the side of the blade, just below the bottom of the cut to raise the bottom of the cut. Or a little solder can be added to the bottom of the cut. Solder is very soft and will not last long. So a duplicate will need to be made from your impressioned key.

If you find that you have lowered a particular cut to the maximum depth without finding the shear line you have filed too far. To save the good part of your work duplicate the blank, except for the one overly-deep cut, then continue impressioning with the duplicate blank.

After you have impressioned one lock in a master key system the other locks will probably have only two or three pins with different depth cuts. If you impression a few different locks you will soon have a master key at some level.

A lock that has been oiled can be extremely hard to impression. Non-residue electronics spray cleaner will probably work well. After flushing out the lock you can speed up the drying by blowing some air into the keyway. There are canned compressed air dusters suitable. Disk tumbler (wafer) locks can be easily impressioned using the same techniques described for pin tumbler locks. However, pull-out techniques should not be used because of possible damage to the discs.

A little less turning pressure is used when impressioning disk locks as compared to pin tumbler locks. The impression marks made by disk locks may look different than the marks made by pin tumbler locks. Depending on exactly how the disk is contacting the blank, they can be anything from a small dot at the edge of the blank, to a straight line across the width of the blank.

Sometimes it is possible to determine the key cut depths for a disk lock without doing any impressioning at all. The technique is called reading the lock. With practice it can be done in seconds. To read a disk lock use a straight pick to lift up all the tumblers. Slowly pull out the pick, watching each tumbler as it falls. You will see that some disks protrude further down into the keyway than others. Typical disk locks use five different depths, numbered number 1 through to number 5. Number 1 cut is at or near to the full height of the key blade, and a number 5 cut being the deepest. The number 1 tumblers protrude the least amount into the keyway and the number 5 cuts protrude the most. By comparing the amount each disc protrudes with respect to the other disc, and with respect to landmarks in the keyway (such as the side warding), it is possible to estimate the depth number of the cut. Usually, the difference between cut depths for disk locks ranges between .015" - .025", with .020" being very common.

Here are some common depths:-

cut number	depth	cut number	depth
1	.240"	4	.180"
2	.220"	5	.160"
3	.200"		

Specific depths for particular locks can be found in depth and spacing manuals, or by taking measurements on keys for other locks of the same type. A disk tumbler lock must be in its shell to be read properly.

LOCKSMITHING TECHNIQUES

Everyone who has been the cinema has seen the agile-fingered thief twirling the dial on a safe then finds the jewels, bonds or cash. Most of these operators astonish the average layman with their ability to open massive safes in practically no time at all.   High-grade safe combination locks are not vulnerable to opening solely by feel, sound or touch. Even a light standard three tumbler dial lock has one million different combinations. While an expert can often open these with little trouble, he must use more mathematics than the average crook ever learned. He must know what the lock on that particular safe is like and how to transpose what he feels or hears to what he knows of the interior of the lock. Needless to say, constant improvements in design and construction are making clean openings difficult even for leading legitimate experts.

There was a college professor of mathematics who bought a heavy bullion safe from a small town bank. The strongbox had been discarded when the bank was refurbished. The cashier who had used it had died while the safe lay empty and locked in the cellar. The lock, a four tumbler one made to withstand attack had the possibility of 100,000,000 possible combinations. With the help of a safe and vault expert, the professor determined one number in the combination (the last one) and thereby reduced to 1,000,000 the combinations he was to try. He and a number of interested students made a list of all the possible combinations, and determined to try each of them. After fifteen months and trying 59,000 combinations, they finally opened the safe. All combination locks depend on the same principles of construction. The amount of security offered depends entirely on the care and precision used in manufacturing. A roughly or loosely made lock naturally will not defy attack as efficiently as a tightly fitted accurately machined one.

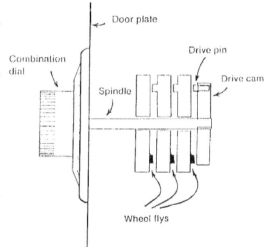

Door plate

Drive pin

Drive cam

Combination dial

Spindle

Wheel flys

The dial face, on which the numbers are stamped, is fixed rigidly on a spindle which penetrates the thickness of the door and is attached to a driving tumbler or flat round wheel parallel to the dial. When the dial is turned this driving tumbler turns with it, and by means of a post inserted in it picks up and rotates a second tumbler, which in turn picks up a third etc. Each of these tumblers has a small notch cut in it. This is just large enough to accommodate the fence which, when all three or four notches are lined up in the proper position, falls into the cavity thus created and thereby actuates the boltwork to release the door.

Dial or combination locks are regularly manufactured with as many as 100,000,000 different combinations, although the average office safe has closer to 1,000,000. Lighter safe locks are available with about 11,000 changes and inexpensive drawer locks with considerably fewer than that. Workmanship and quality of materials being equal, the security of a dial lock depends to a great extent upon the number of possible combinations to which it is adaptable. A lock with three tumblers, each tumbler being graduated for one hundred different fence slot positions has 1,000,000 combinations. This is obtained by raising the 100 to the 3rd power. A four tumbler lock of similar type would have 100,000,000 possibilities for changes, etc.

Service work of any kind should never be attempted on combination locks by anyone not thoroughly acquainted with this branch of locksmithing. This caution cannot be over-emphasised. The key changing lock was devised to eliminate the necessity of disassembling a lock to change the combination. On these locks, a special combination key is employed to unlock the tumblers from the spindle.

The new combination is set to whatever numbers are desired while the changing key remains in the lock. After the key is turned back a half or a quarter turn to lock the tumblers in place, the safe should open on the combination to which it was set. Never close or lock a safe door after a combination change until the new combination has been tried and actually operated at least three times without a single failure. Disregarding this rule may easily result in a lockout.

The great advantage of the key changing combination lock is that it relieves the locksmith who changes the combination of any suspicion if some unauthorised person gets the combination. On this type of lock the safe expert works at the back of the door, releasing the tumblers with the combination key. A locksmith engaged in safe work accepts a great responsibility. Besides the inconvenience to the customer, it is the danger to his own reputation. A lockout on a safe or vault door can be very costly if an expert must be brought in to rectify a mistake.

Most new safes have combination instruction sheets attached to them. These should be read carefully for each lock, and never disregarded.

There are two general types of combination locks:-

- hand changing
- key changing

Although hand changing is the simpler, it is in this kind that mistakes are most often made since the tumblers must be removed from the lock to be adjusted. Be absolutely certain to return all the tumblers, spring rings and tension discs to the locks in exactly the same relative positions as they were originally. Never lubricate a combination lock. Tension washers in these most intricate of mechanically operated devices are set to operate on friction alone. If the friction is decreased the lock may fail to operate properly.

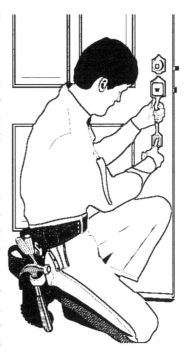

Practically all combination lock manufacturers request that locksmiths do not attempt to make repairs in the field. This is not only for the customer's and the manufacturer's protection but for the protection of the locksmith involved.

A locksmith called to repair a burglarised safe should always install a new lock, not repair the old one. Most insurance companies insist on this to avoid future claims if a further burglary takes place and the lock was defective from the previous battering.

Protecting Safe Locks from Attack

Since combination locks are built primarily for accuracy and combination security, they must be attached to the safe doors in such a way as to be protected as much as possible from forcible attack.

Many modern burglar resistant safes are made with what is known as a drive proof spindle, precluding the possibility of knocking off the exterior combination dial and driving the spindle and tumblers into the safe. This method of forcible entry is known as punching. The drive proof spindle is larger at the front of the door than it is at the inside, preventing its being driven through the door.

Another often used is a lock with an indirect or geared drive. In this type of construction instead of the lock tumblers being directly on the end of the spindle, a simple toothed gear takes their place. This gear meshes with another gear attached to the drive tumbler which is located in the lock case either to the right or left, or above or below the dial.

Another measure that decreases the possibilities of a successful attack by a criminal is the automatic relocking device. Almost all recently manufactured safe lock is now equipped with this system. It usually consists of an auxiliary mechanism which jams the boltwork, even if the lock itself is rendered useless. Therefore a safecracker who destroyed the combination lock would only succeed in locking himself out more securely. A relocking mechanism is not constructed so as to be operated from the outside, except by factory trained service experts equipped with special tools.

Time Locks

Time locks operate exclusively by clockwork, depending neither on combination or keys for control. Usually at least three clocks are set in a series, any one of which will individually operate the lock. So if one stops and one is put out of service by an accident, such as a slight earthquake, there is still one left to actuate the lock.

They are known as 72 or 96 hour clocks, depending upon the number of hours their opening may be delayed. Locks of this type should never, under any circumstances, be touched by anyone not entirely familiar with their operation. Repairs and adjustments are generally made by factory experts, usually on a contract basis. These men are not only master locksmiths but must also be qualified jewellers or watchmakers to properly adjust, repair and clean the delicate clockwork.

Vault Locks

Safe-deposit and currency vaults in present day banks are almost universally controlled by a combination of two types of locks.

Dial locks are installed in conjunction with the time lock, so that the vault can be opened only when certain authorised bank officials are present to operate the combination. The time lock for its contribution to security, keeps these or officials or anyone else from opening the vault at night or during holidays. This plan precludes the possibility of criminals kidnapping a bank employee and forcing them to open the locked vault. Signs in many cases are publicly displayed in banks to notify all persons of this double protection, thus saving employees from possible mistreatment or torture by armed robbers. Another plan for increasing the security of a vault is to distribute the combination to three or four persons. This makes it necessary for all to be present at every opening, since each knows only his part of the combination. This is known as the multiple custody system and relieves an individual employee of a great deal of worry and responsibility.

The largest users of the double custody principle are the safe deposit companies. Locks used on safe deposit boxes differ from any other locks in two ways. First they require two separate and distinct keys to operate them, that of the renter or user, and that of the guard or bank attendant. Both must be used, simultaneously in most cases, to open each individual box. Safe deposit box locks are not made so that any master key or pass key will fit any of them. Each lever lock of this type has from five to eight or ten tumblers depending on the size of the vault, each tumbler having five or six more different depth possibilities. A fair average would be six and six, or 46,656 different key changes for a light six lever lock. Since every bank's guard or preparatory key is different and each vault has no two locks keyed alike, there is little danger of unauthorised entry. This is since most safe deposit firms require signature identification before the renter is even admitted to the vault. One modern innovation in bank and safe locks is the timbination lock. Essentially it consists of a combination lock with a double movement timing arrangement to delay opening of the lock for from five to thirty minutes. It is intended for use where such a feature is desired to prevent daylight robberies or forced operation under threat. Assume for instance that a jeweller finds an armed robber waiting for him when he opens the shop in the morning. Forced at gun point to open the safe, he does so. But the timing device keeps the door closed for thirty minutes after the lock is opened. The thief will probably leave rather than wait half an hour, since he knows a secret alarm may have been activated and that other employees and customers will soon be entering the building.

Many locksmiths have regular customers among business firms who pay top prices for insurance and legal advice. Quite often a locksmith can serve in the capacity of an adviser and be well paid. Probably many of the following rules for security of the contents of a safe are disregarded in most offices. Locksmiths who notice these possibilities for loss can command excellent fees for suggestions to correct them.

- Never write a safe combination anywhere an unauthorised person can get at it. Write it on a card and put it in your safe deposit box for your own protection in case of loss of memory or death.

- Change the combination without fail immediately upon discharging a person to whom you have given the combination, or if you have reason to believe someone has acquired it without authorisation.

- Always stand in front of a safe when operating the combination. Standing to the side to get better light may give the combination to anyone near by. Safe robbers have even been known to watch with field glasses from windows across the street.

- Never store cash or valuables in a safe intended solely for fire protection. These boxes are not intended to be burglar proof, and in many cases are not even burglar resistant. Anchor a burglar proof chest to the bottom of your fire safe if you customarily store large amounts of cash overnight.

- Never place a safe containing cash or articles of great value against a back or side wall. Better to have it near the front of the store where it can be easily seen from the street. Too often a jeweller has opened his safe to find the contents missing, the thieves having entered an adjoining building and cut the back of the safe away without disturbing his intricate burglar alarms.

- Do not have safe combinations set to match numbers in your house address, telephone numbers or licence plates.

Other people have phone directories too.!

GLOSSARY

A

actuator

A device, usually connected to a cylinder, which, when activated, may cause a lock mechanism to operate.

adjustable mortise cylinder

Any mortise cylinder whose length can be adjusted for a better fit in doors of varying thickness.

angle of cut

Cut angle.

angularly bitted key

A key which has cuts made into the blade at various degrees of rotation from the perpendicular.

armoured front

Face plate.

associated change key

A change key directly related to a particular master key(s), through the use of constant cuts.

associated master key

A master key which has particular change keys that relate directly to its combination through the use of constant cuts.

B

back of blade

Bottom of blade.

back plate

A thin piece of metal, usually with a concave portion, which is used with machine screws to fasten certain types of cylinders to a door.

backed off blade

Radiused blade bottom.

ball bearing

1. A metal ball used in the pin stack to accomplish some types of hotel or construction keying.
2. A ball, usually made of steel, used by some lock manufacturers as the bottom element in the pin stack in one or more pin chambers.
3. Any metal ball used as a tumbler's primary component.

BLI

British Locksmith Institute.

ball end pin

Bottom pin.

barrel

Cylinder plug.

Bell type key

A key whose cuts are in the form of wavy grooves milled into the flat sides of the key blade. The grooves usually run the entire length of the blade.

bible

The portion of the cylinder shell which houses the pin chambers, especially those of a key-in-knob cylinder or certain rim cylinders.

bicentric cylinder

A cylinder which has two independent plugs, usually with different keyways. Both plugs are operable from the same face of the cylinder. It is designed for use in extensive master key systems.

bi-directional cylinder

A cylinder which may be operated clockwise and anticlockwise by a single key.

binary cut key

A key whose combination only allows for two possibilities in each bitting position: cut/no cut.

binary type cylinder or lock

A cylinder or lock whose combination only allows for two bitting possibilities in each bitting position.

bit

1. The part of the key which serves as the blade, usually in a warded or lever tumbler lock.
2. See key cut(s).

bitting

1. The numbers which represent the dimensions of the key cut(s).
2. The actual cut(s) or combination of a key.

bitting depth

The depth of a cut made in the blade of a key.

bitting list

A list of all the key combinations used within a system. The combinations are usually arranged in order of the blind code, direct code, and/or key symbol.

bitting position

The location of a key cut.

blade

The portion of a key which may contain the cuts and/or milling.

blade tumbler

See "disc tumbler".

blank

1. Key blank.
2. Uncut

blind code

A designation, unrelated to the bitting, assigned to a particular key combination for future reference when additional keys or cylinders may be needed.

block master key

The one pin master key for all combinations listed as a block in the standard progression format.

blocking ring

See "cylinder collar".

blockout key

See "lockout key".

bottom of blade

The portion of the blade opposite the cut edge of a single bitted key.

bottom pin

Usually a cylindrical shaped tumbler which may be conical, ball shaped or chisel pointed on the end which makes contact with the key.

bow

The portion of the key which serves as a grip or handle.

bow stop

A type of stop located near the key bow.

broach

A tool used to cut the keyway into the cylinder plug.

builders' master key.

See "construction master key"

building master key

A master key used to operate all locks in a given building.

bypass key

The key which operates a key override cylinder.

C

cam

A flat actuator or locking bolt attached to the rear of a cylinder perpendicular to its plug and rotated by the key.

cam lock

A complete locking assembly in the form of a cylinder whose cam is the actual locking bolt.

cap

1. A spring cover for a single pin chamber.
2. A part which may serve as a plug retainer and/or a holder for the tailpiece.

capping block

A holding fixture for certain interchangeable cores which aids in the installation of the caps.

cell

See "pin chamber".

central key system

See "maison key system".

chain key system

See "selective key system".

chamber

Any cavity in a cylinder plug and/or shell which houses the tumbler(s).

Change key

1. A key which operates only one cylinder or one group of keyed alike cylinders in a keying system.
2. See "reset key".

change key constant

See "constant cut".

change key section

See "single key section".

changeable bit key

A key which can be re-combined by exchanging and/or rearranging portions of its bit or blade.

chip

see "master pin".

clutch

The part of the profile cylinder which transfers rotational motion from the inside or outside element to a common cam or actuator.

code

1. A designation assigned to a particular key combination for reference when additional keys or cylinders may be needed. See also, "blind code", "direct code"

and "key symbol".

2. See "combinate".

code key

A key cut to a specific code rather than duplicated from a pattern key. It may or may not conform to the lock manufacturer's specifications.

code list

1. See "bitting list".
2. See "progression list".

code machine

See "key coding machine".

code number

1. See "blind code".
2. See "direct code".

code original key

A code key which conforms to the lock manufacturer's specifications.

column master key

See "vertical group master key".

combinate

To set a combination in a lock, cylinder or key.

combination

The group of numbers which represent the bitting of a key and/or the tumblers of a lock or cylinder.

combination wafer

Atype of disc tumbler used in certain binary type disc tumbler key-in-knob locks. Its presence requires that a cut be made in that position of the operating keys.

common keyed

See "maison key system.

compensate drivers

1. To select longer or shorter top pins, depending on the length of the rest of the pin stack, in order to achieve a uniform pin stack height.
2. See "graduated drivers".

complementary keyway

Usually a disc tumbler keyway used in master keying. It accepts keys of different sections whose blades contact different bearing surfaces of the tumblers.

composite keyway

A keyway which has been enlarged to accept more than one key section, often key sections of more than one manufacturer.

concealed shell cylinder

A specially constructed (usually mortise) cylinder. Only the plug face is visible when the lock trim is in place.

connecting bar

See "tailpiece".

constant cut

1. Any bitting(s) which are identical in corresponding positions from one key to

another in a keying system. They usually serve to group these keys together within a given level of keying, and/or link them with keys of other levels.

2. See "rotating constant".

construction breakout key

A key used by some manufacturers to render all construction master keys permanently inoperative.

construction core

An interchangeable or removable core designed for use during the construction phase of a building. The cores are normally keyed alike and, upon completion of construction, they are to be replaced by the permanent system's cores.

construction master key

Akey normally used by construction personnel for a temporary period during building

construction. It may be rendered permanently inoperative without disassembling the cylinder.

eanstruclion master keyed

Of or pertaining to a cylinder which is or is to be operated temporarily by a construction master key.

control cut

Any bitting which operates the retaining device of an interchangeable or removable core.

control key

A key whose only purpose is to remove and/or install an interchangeable or removable core.

control lug

the part of an interchangeable or removable core retaining device which locks the core into its housing.

control sleeve

The part of an interchangeable core retaining device which surrounds the plug.

controlled cross keying

A condition in which two or more different keys of the same level of keying and under the same higher level key(s) operate one cylinder by design.

core

A complete unit often with a figure eight shape which usually consists of the plug, shell, tumblers, springs, plug retainer and spring cover(s). It is primarily used in removable and interchangeable core cylinders and locks.

cross keying

The deliberate process of combining a cylinder (usually in a master key system) to two or more different keys which would not normally be expected to operate it together.

cut

1. See "key cut(s)".
2. To make cuts into a key blade.

cut angle

A measurement usually expressed in degrees of the angle between the two sides of a key cut.

cut depth

See "bitting depth".

cut edge

The portion of the key blade which contains the cuts.

cut key

A key which has been bitted or combined.

cut profile

See "key cut profile".

cut root

The bottom of a key cut.

cut root shape

The shape of the bottom of a key cut. It may have a flat or radius of a specific dimension, or be a perfect "V".

cut rotation

See "degree of rotation".

cutter

The part of a key machine which makes the cuts into the key blank.

cylinder

A complete operating unit which usually consists of the plug, shell, tumblers, springs, plug retainer, a cam/tailpiece or other actuating device, and all other necessary operating parts.

cylinder assembly

See "cylinder".

cylinder bar

See "tailpiece".

cylinder blank

A dummy cylinder which has a solid face and no operating parts.

cylinder clip

A spring steel device used to secure some types of cylinders.

cylinder collar

A plate or ring installed under the head of a cylinder to improve appearance and/or security.

cylinder guard

A protective cylinder mounting device.

cylinder key

A broad generic term including virtually all pin and disc tumbler keys.

cylinder plug

See "plug".

cylinder ring

See "cylinder collar".

cylinder rage

See "cylinder collar".

cylinder shell

See "shell".

D

day key

See "change key".

declining step key

A key whose cuts are progressively deeper from bow to tip.

decode

To determine a key combination by physical measurement of a key and/or cylinder parts.

degree of rotation

A specification for the angle at which a cut is made into a key blade as referenced from the perpendicular. This specification is typically used for some high security keys.

department master key

A master key which accesses all areas of a given department.

depth

See "bitting depth".

depth key set

A set of keys used to make a code original key on a key duplicating machine to a lock manufacturer's given set of key bitting specifications. Each key is cut with the correct spacing to one depth only in all bitting positions, with one key for each depth.

derived series

A series of blind codes and bittings which are directly related to those of another bitting list.

detainer disc

See "rotary tumbler".

dimple

A key cut in a dimple key.

dimple key

A key whose cuts are drilled or milled into its blade surfaces. The cuts normally do not change the blade silhouette.

direct code

A designation assigned to a particular key which includes the actual combination of the key.

disc

1. See "disc tumbler".

2. See "master pin".

3. See "rotary tumbler".

Disc tumbler

1. A flat tumbler which must be drawn into the cylinder plug by the proper key so that none of its extremities extends into the shell.

2. A flat usually rectangular tumbler with a gate which must be aligned with a sidebar by the proper key.

display key

A special change key in a hotel master key system which will allow access to one designated guest room, even if the lock is in the shut out mode. It may also act as a shut out key for that room.

double bitted key

A key bitted on two opposite surfaces.

double pin

To place more than one master pin in a single pin chamber.

double sided key

See "double bitted key".

driver

See "top pin".

driver spring

A spring placed on top of the pin stack to exert pressure on the pin tumblers.

drop

1. See "increment".

2. A pivoting or swinging dust cover.

dummy cylinder

A non-functional facsimile of a rim or mortise cylinder used for appearance only, usually to conceal a cylinder hole.

duplicate

1.   See "duplicate key".

2.   To copy.

3.

duplicate blank

See "non-original key blank".

duplicate key

Any key reproduced from a pattern key.

dust cover

A device designed to prevent foreign matter from entering a mechanism through the keyway.

dustproof cylinder

A cylinder designed to prevent foreign matter from entering either end of the keyway.

E

effective plug diameter

The dimension obtained by adding the root depth of a key cut to the length of its correspond¬ing bottom pin which establishes a perfect shear line.

ejector hole

A hole found on the bottom of certain interchangeable cores under each pin chamber. It provides a path for the ejector pin.

ejector pin

A tool used to drive all the elements of a pin chamber out of certain interchangeable cores.

emergency key

1. See "emergency master key".

2. The key with operates a privacy function lockset.

emergency master key

A special master key which usually operates all guest room locks in a hotel master key system at all times, even in the shut out mode. This key may also act as a shut out key.

EMK

Emergency master key.

encode

See "combinate".

ENG

Symbol for engineer's key.

engineer's key

A selective master key which is used by maintenance personnel to operate many locks under different master keys in the system of three or more levels of keying.

escutcheon

A surface mounted trim which enhances the appearance and/or security of a lock installation.

extractor key

A tool which normally removes a portion of a two-piece key or blocking device from a keyway.

F

face plate

A mortise lock cover plate exposed in the edge of the door.

factory original key

The cut key furnished by the lock manufacturer for a lock or cylinder.

false plug

See "plug follower".

fence

1. A projection on a lock bolt which prevents movement of the bolt unless it can enter gates of properly aligned tumblers.

2. See "sidebar".

file key

See "pattern key".

finish

A material, colouring and/or texturing specification.

fireman's key

A key used to override normal operation of lifts, bringing them to the ground floor.

first generation duplicate

A key which was duplicated using a factory original key or a code original key as a pattern.

first key

Any key produced without the use of a pattern key.

five column progression

A process wherein key bittings are obtained by using the cut possibilities in five columns of the key bitting array.

five pin master key

A master key for all combinations obtained by progressing five bitting positions.

flexible head mortise cylinder

An adjustable mortise cylinder which can be extended against spring pressure to a slightly longer length.

floating master key

1. See "unassociated master key".

2. See "selective master key".

floor master key

A master key which operates all or most cylinders on a particular floor of a building.

follower

See "plug follower".

formula

See "key bitting array".

four column progression

A process where key bittings are obtained by using the cut possibilities in four columns of the key bitting array.

four pin master key

A master key for all combinations obtained by progressing four bitting positions.

G

genuine key blank

See "original key blank

ghost key

See "incidental master key".

GM

Grand master key.

graduated drivers

1. A set of top pins of different lengths. Usage is based on the height of the rest of the pin stack, in order to achieve a uniform pin stack height.

2. See "compensate drivers".

grand master key

The key which operates two or more separate groups of locks, which are each operated by a different master key.

grand master key system

A master key system which has exactly three levels of keying.

grand master keyed

Of or pertaining to a lock or cylinder which is or is to be keyed into a grand master key system.

guide keys

See "depth key set".

H

hardware schedule

A listing of the door hardware used on a particular job. It includes the types of hardware, manufacturers, locations, finishes and sizes. It should include a keying schedule specifying how each locking device is to be keyed.

HGNI

Horizontal group master key.

high security cylinder

A cylinder which offers a greater degree of resistance to any or all of the following: picking, impressioning, key duplication, drilling or other forms of forcible entry.

high security key

A key for a high security cylinder.

hold and vary

See "rotating constant method".

hold open cylinder

A cylinder provided with a special cam which will hold a latch bolt in the retracted position.

hollow driver

A top pin hollowed out on one end to receive the spring, used in cylinders with extremely limited clearance in the pin chambers.

horizontal group master key

The two pin master key for all combinations listed in all blocks in a line across the page in the standard progression format.

housekeeper's key

A selective master key in a hotel master key system which may operate all guest and linen rooms and other housekeeping areas.

housing

The part of a locking device which is designed to hold a core.

I

imitation blank

See "non-original key blank".

impression

1.   The mark made by a tumbler on its key cut.

2.   To fit a key by the impression technique.

impression technique

The means of fitting a key directly to a locked cylinder, by manipulating a blank in the keyway and cutting the blank where the tumblers have made marks.

incidental master key

A key cut to an unplanned shearline created when the cylinder is combinated to the top master key and a change key.

increment

A usually uniform increase or decrease in the successive depths of a key cut which must be matched by a corresponding change in the tumblers.

indicator

A device which provides visual evidence that a deadbolt is extended or that a lock is in the shut out mode.

indirect code

See "blind code".

individual key

An operating key for a lock or cylinder which is not part of a keying system.

interchange

See "key interchange".

interchangeable core

A key removable core which can be used in all, or most, of the core manufacturer's product line. No tools, other than the control key, are required for removal of the core.

interlocking pin tumbler

A type of pin tumbler which is designed to be linked together with all other tumblers in its chamber when the cylinder plug is in the locked position.

J

jiggle key

See "manipulation key".

jumbo cylinder

A rim or mortise cylinder.

K

k

The symbol used for "keys" used after a numerical designation of the quantity of the keys requested to be supplied. (e.g., lk, 2k, 3k, etc.)

KA

Keyed alike.

KRA

Key bitting array.

key

A properly combinated device which is, or most closely resembles, the device specifically intended by the lock manufacturer to operate the corresponding lock.

key bitting punch

A manually operated device which stamps or punches the cuts.

key blank

Any material manufactured to the proper configuration which allows its entry into the keyway of a specific locking device.

key coding machine

A key machine designed for the production of code keys. It may or may not also serve as a duplicating machine.

key control

Any method or procedure which limits unauthorised acquisition of a key and/or controls distribution of authorised keys.

key cut(s)

The portion of the key blade which remains after being cut and which aligns the tumblers.

key cut profile

The shape of a key cut.

key duplicating machine

A key machine designed to make copies from a pattern key.

key gauge

A usually flat device with a cutaway portion indexed with a given set of depth or spacing specifications. It is used to help determine the combination of a key.

key-in-knob cylinder

A cylinder used in a key-in-knob lockset.

key interchange

An undesirable condition, usually in a master key system, whereby a key unintentionally operates a cylinder or lock.

key machine

Any machine designed to cut keys.

key manipulation

Manipulation of an incorrect key in order to operate a lock or cylinder.

key override

A provision allowing interruption or circumvention of normal operation of a combination lock or electrical device.

key override cylinder

A lock cylinder installed in a device to provide a key override function.

key picking

See "key manipulation".

key pin

See "bottom pin".

key profile

See "key section".

key pull position

Any position of the cylinder plug at which the key can be removed.

key punch

See "key bitting punch".

key records

Records which typically include some or all of the following: bitting list, key bitting array, key system schematic, end user, number of keys/cylinders issued, names of persons to whom key-, were issued, hardware/keying schedule.

key section

The exact cross sectional configuration of a key blade as viewed from the bow toward the tip.

key stop

See "stop (of a key)".

key symbol

A designation used for a key combination in the standard key coding system.

key system schematic

A drawing with blocks using keying symbols.

key trap core/cylinder

A special core or cylinder designed to capture any key to which it is combinated, once that key is inserted and turned slightly.

keyed

Having provision for operation by key.

keyed common

See "maison key system".

keying

Any specification for how a cylinder or group of cylinders are or are to be combinated in order to control access.

keying chart

See "pinning chart".

keying diagram

See "key system schematic".

keying kit

A compartmented container which holds an assortment of tumblers, springs and/or other parts.

keying levels

See "levels of keying".

keying schedule

n. a detailed specification of the keying system listing how all cylinders are to be keyed and the quantities, markings, and shipping instructions of all keys and/or cylinders to be provided.

keyway

1. The opening in a lock or cylinder which is shaped to accept a key bit or blade of a proper configuration.

2. The exact cross sectional configuration of a keyway as viewed from the front. It is not necessarily the same as the key section.

keyway shutter

See "dust cover".

keyway unit

The plug of certain binary type disc tumbler key-in-knob locks.

KR

1.   Keyed random.

2.   Key retaining.

L

layout board

See "layout tray".

layout tray

A compartmented container used to organise cylinder parts during keying or servicing.

1.

levels of keying

The divisions of a master key system into hierarchies of access, as shown in the following tables. Note that the standard key coding system has been expanded to include key symbols for systems of more than four levels of keying.

TWO LEVEL SYSTEM

level of keying	key name	abb.	key symbol
Level II	master key	MK	AA
Level I	change key	CK	1AA, 2AA, etc.

THREE LEVEL SYSTEM

level of keying	key name	abb.	key symbol
Level HI	grand master key	GMK	A
Level II	master key	MK	AA, AB, etc.
Level I	change key	CK	AA1, AA2, etc.

FOUR LEVEL SYSTEM

level of keying	key name	abb.	key symbol
Level IV	great grand master key	GGMK	GGMK
Level III	grand master key	GMK	A, B, etc.
Level II	master key	MK	AA, AB, etc.
Level I	change key	CK	AA1, AA2, etc.

FIVE LEVEL SYSTEM

level of keying	key name	abb.	key symbol
Level V	great great grand master key	GGGMK	GGGMK
Level IV	great grand master key	GGMK	A, B, etc.
Level III	grand master key	GMK	AA, AB, etc.
Level II	master key	MK	AAA, AAB, etc.
Level I	change key	CK	AAA1, AAA2, etc.

SIX LEVEL SYSTEM

level of keying

Level VI Level V Level IV Level III Level II Level I

key name	abb.	key symbol
great great grand master key	GGGMK	GGGIVIK
great grand master key	GGMK	A, B, etc.
grand master key	GMK	AA, AB, etc.
master key	MK	AAA, AAB, etc.
sub-master key	SMK	AAAA, AAAB, etc.
change key	CK	AAAA1, AAAA2, etc.

lever tumbler

A flat, spring-loaded tumbler which pivots on a post. It contains a gate which must be aligned with a fence to allow movement of the bolt.

loading tool

A tool which aids installation of cylinder components into the cylinder shell.

lockout

Any situation in which the normal operation of a lock or cylinder is prevented.

lockout key

A key made in two pieces. One piece is trapped in the keyway by the tumblers when inserted and blocks entry of any regular key. The second piece is used to remove the first piece.

M

MACS

Maximum adjacent cut specification.

maid's master key

The master key in a hotel master key system given to the maid. It operates only cylinders of the guest rooms and linen closets in the maid's designated area.

maintenance master key

See "engineer's key".

maison key system

From the French meaning "house", a keying system in which one or more

cylinders are operated by every key.

manipulation key

Any key other than a correct key which can be variably positioned and/or manipulated in a keyway to operate a lock or cylinder.

master

See "master key".

master blank

See "multi-section key blank".

master chip

See "master pin".

master disc

See "master pin".

master key

1.   A key which operates all master keyed locks or cylinders in a group, each lock or cylinder usually operated by its own change key.

2.   To combinate a group of locks or cylinders such that each is operated by its own change key as well as by a master key for the entire group.

master key changes

The number of different usable change keys available under a given master key.

master key system

Any keying arrangement which has two or more levels.

master lever

A lever tumbler which can align some or all other levers in its lock so that their gates are at the fence. It is typically used in locker locks.

master pin

1. Usually a cylindrical shaped tumbler, flat on both ends, placed between the top and bottom pin to create an additional shear line.

2. A pin tumbler with multiple gates to accept a sidebar.

master ring

A tube shaped sleeve located between the plug and shell of certain cylinders to create a second shear line.

master ring lock cylinder

A lock or cylinder equipped with a master ring.

maximum adjacent cut differential

See "maximum adjacent cut specification".

maximum adjacent cut specification

The maximum allowable difference between adjacent cut depths.

maximum opposing cut specification

The maximum allowable depths to which opposing cuts can be made without breaking through the key blade. This is typically a consideration with dimple keys.

MK

Master key.

MK'd

Master keyed.

MK'd only

Master keyed only.

MK section

Master key section.

MLA

Master Locksmithing Association.

MOCS

Maximum opposing cut specification.

mogul cylinder

A very large pin tumbler cylinder whose pins, springs, key, etc. are also proportionally increased in size. It is typically used in prison locks.

mortise cylinder

A threaded cylinder typically used in mortise locks of American manufacture.

mortise cylinder blank

See "cylinder blank".

movable constant

See "rotating constant".

multi-section key blank

A key section which enters more than one, but not all keyways in a multiplex key system.

multiple gating

The means of master keying by providing a tumbler with more than one gate.

multiplex key blank

Any key blank which is part of a multiplex key system.

mushroom driver

See "mushroom pin".

mushroom pin

A pin tumbler, usually a top pin, which resembles a mushroom. It is typically used to increase pick resistance.

N

NCK

Symbol for "no change key," pimarily used in hardware schedules.

negative locking

Locking achieved solely by spring pressure or gravity which prevents a key cut too deeply from operating a lock or cylinder.

O

one pin master key

A master key for all combinations obtained by progressing only one bitting position.

open code

See "direct code".

operating key

Any key which will properly operate a lock or cylinder to lock or unlock the lock mechanism and is not a control key or reset key.

original key

See "factory original key".

original key blank

A key blank supplied by the lock manufacturer to fit that manufacturer's specific product.

P

pattern key

An original key kept on file to use in a key duplicating machine when additional keys are required.

phantom key

Incidental master key.

Pick

A tool or instrument, other than the specifically designed key, made for the purpose of manipulating tumblers in a lock or cylinder into the locked or unlocked position through the keyway, without obvious damage.

pick key

A type of manipulation key, cut or modified to operate a lock or cylinder.

pin cell

See "pin chamber".

pin kit

A type of keying kit for a pin tumbler mechanism.

pin segment

See "pin tumbler".

pin stack

All the tumblers in a given pin chamber.

pin stack height

The measurement of a pin stack, often expressed in units of the lock manufacturer's increment or as an actual dimension.

pin tray

See "layout tray".

pin tumbler

Usually a cylindrical shaped tumbler. Three types are normally used: bottom pin, master pin and top pin.

pin tweezers

A tool used in handling tumblers and springs.

pinning block

A holding fixture which assists in the loading of tumblers into a cylinder or cylinder plug.

pinning chart

A numerical diagram which indicates the sizes and order of installation of the various pins into a cylinder.

plug

The part of a cylinder which contains the keyway, with tumbler chambers usually corresponding to those in the cylinder shell.

plug follower

A tool used to allow removal of the cylinder plug while retaining the top pins, springs, and/or other components within the shell.

plug holder

A holding fixture which assists in the loading of tumblers into a cylinder plug.

plug retainer

The cylinder component which secures the plug in the shell.

plug set-up chart

See "pinning chart".

plug vice

See "plug holder".

positional master keying

A method of master keying typical of certain binary type disc tumbler key-in-knob locks and of magnetic and dimple key cylinders

.

positive locking

The condition when a key cut which is too high forces its tumbler into the locking position.

practical key changes

The total number of usable different combinations available for a specific cylinder or lock mechanism.

prep key

A type of guard key for a safe deposit box lock with only one keyway. It must be turned once and withdrawn before the renter's key will unlock the unit.

profile

See "key section".

profile cylinder

A cylinder with a usually uniform cross section, which slides into place and usually is held by a mounting screw. It is typically used in mortise locks of non-U.S. manufacture.

progress

To select possible key bittings from the key bitting array, usually in numerical order.

progression

A logical sequence of selecting possible key bittings, usually in numerical order from the key bitting array.

progressive

Any bitting position which is progressed rather than held constant.

proprietary

A keyway and key section assigned exclusively to one end user by the lock manufacturer. It may also be protected by law from duplication.

R

radiused blade bottom

The bottom of a key blade which has been radiused to conform to the curvature of the cylinder plug it is designed to enter.

random master keying

Any undesirable process used to master key which uses unrelated keys to create a system.

rap

To unlock a plug from its shell by striking sharp blows to the spring of the cylinder while applying tension to the plug.

read key

A key which allows access to the sales and/or customer data on certain types of cash control equipment (e.g., cash registers).

recode

See "recombinate".

recombinate

To change the combination of a lock, cylinder, or key.

recore

To rekey by installing a different core.

register groove

The reference point on the key blade from which some manufacturers locate the bitting depths.

register number

A reference number, typically assigned by the lock manufacturer.

rekey

To change the existing combination of a cylinder or lock.

removable cylinder

A cylinder which can be removed from a locking device by a key and/or tool.

removal key

The part of a two-piece key which is used to remove its counterpart from a keyway.

renter's key

A key which must be used together with a guard key, prep key or electronic release to unlock a safe deposit lock. It is usually different for every unit within an installation.

repin

To replace pin tumblers, with or without changing the existing combination.

reserved

See "restricted".

reset

See "recombinate".

reset key

A key used to set some types of cylinders to a new combination. Many of these cylinders require the additional use of tools and/or the new operating key to establish the new combination.

restricted

A keyway and corresponding key blank whose sale and/or distribution is limited by the lock manufacturer in order to reduce unauthorised key proliferation.

reversible key

Asymmetrical key which may be inserted either way up to operate a lock.

rim cylinder

A cylinder typically used with surface applied locks and attached with a back plate and machine screws. It has a tailpiece to actuate the lock mechanism.

rocker key

Manipulation key.

root depth

The dimension from the bottom of a cut on a key to the bottom of the blade.

root of cut

Cut root.

rose

Usually a circular escutcheon.

rotary tumbler

A circular tumbler with one or more gates. Rotation of the proper key aligns the tumbler gates at a sidebar, fence or shackle slot.

rotating constant

One or more cuts in a key of any level which remain constant throughout all levels and are identical to the top master key cuts in their corresponding positions. The positions where the top master key cuts are held constant may be moved, always in a logical sequence.

rotating constant method

A method used to progress key bittings in a master key system, where at least one cut in each key is identical to the corresponding cut in the top master key. The identical cut is moved to different locations in a logical sequence until each possible planned position has been used.

row master key

The one pin master key for all combinations listed on the same line across a page in the standard progression format.

S

S/A

Sub-assembled.

safety factor

See "maximum adjacent cut specification".

sample key

Pattern key.

scalp

A thin piece of metal which is usually crimped or spun onto the front of a cylinder. It determines the cylinder's finish and may also serve as the plug retainer.

schematic

Key system schematic.

second generation duplicate

A key reproduced from a first generation duplicate.

sectional key blank

Multiplex key blank.

sectional keyway system

Multiplex key system.

security collar

A protective cylinder collar.

segmented follower

A plug follower which is sliced into sections which are introduced into the cylinder shell one at a time. It is typically used with profile cylinders.

selective key system

A key system in which every key has the capability of being a master key. It is normally used for applications requiring a limited number of keys and extensive cross keying.

series wafer

A type of disc tumbler used in certain binary type key-in-knob locks.

set-up key

A key used to calibrate some types of key machines.

set-up plug

A type of loading tool shaped like a plug follower. It contains pin chambers and is used with a shove knife to load springs and top pins into a cylinder shell.

seven column progression

A process where key bittings are obtained by using the cut possibilities in seven columns of the key bitting array.

shear line

A location in a cylinder at which specific tumbler surfaces must be aligned, removing obstruc¬tion(s) which prevented the plug from moving.

shedding key

Declining step key.

shell

The part of the cylinder which surrounds the plug and which usually contains tumbler chambers corresponding to those in the plug.

Shim

A thin piece of material used to unlock the cylinder plug from the shell by separating the pin tumbers at the shear line, one at a time.

shoulder

Any key stop other than a tip stop.

shouldered pin

A bottom pin whose diameter is larger at the flat end to limit its penetration into a counter-bored chamber.

shove knife

A tool used with a set-up plug which pushes the springs and pin tumblers into the cylinder shell.

shut out key

Usually used in hotel keying systems, a key that will make the lock inoperable to all other keys in the system except the emergency master key, display key, and some types of shut out keys.

sidebar

A primary or secondary locking device in a cylinder.

simplex key section

A single independent key section which cannot be used in a multiplex key system.

six column progression

A process where key bittings are obtained by using the cut possibilities in six columns of the key bitting array.

six pin master key

A master key for all combinations obtained by progressing six bitting positions.

skew

Rotation.

slide

Spring cover.

SMK

Sub-master key.

spacing

The dimensions from the stop to the center of the first cut and/or to the centres of successive cuts.

special application cylinder

Any cylinder other than a mortise, rim, key-in-knob or profile cylinder.

split pin

See "master pin".

split pin master keying

A method of master keying a pin tumbler cylinder.

spool pin

Usually a top pin which resembles a spool.

standard key coding system

An industry standard and uniform method of designating all keys and/or cylinders in a master key system.

step

Increment.

step pin

A spool or mushroom pin which has had a portion of its end machined to a smaller diameter than the opposite end.

step tolerance

Maximum adjacent cut specification.

stepped tumbler

A special (usually disc) tumbler used in master keying. It has multiple bearing surfaces for blades of different key sections.

stop (of a key)

The part of a key from which all cuts are indexed and which determines how far the key enters the keyway.

sub-assembled

Uncombinated.

sub-master key

The master key level immediately below the master key in a system of six or more levels of keying.

T

tailpiece

An actuator attached to the rear of the cylinder, parallel to the plug, typically used on rim, key-in-knob or special purpose cylinders.

theoretical key changes

The total possible number of different combinations available for a specific cylinder or lock mechanism.

thimble

Plug holder.

threaded cylinder

See "mortise cylinder".

three column progression

A process where key bittings are obtained by using the cut possibilities in three columns of the key bitting array.

three pin master key

A master key for all combinations obtained by progressing three bitting positions.

thumb turn cylinder

A cylinder with a turn knob rather than a keyway and tumbler mechanism.

tip

The portion of the key which enters the keyway first.

tip stop

A type of stop located at or near the tip of the key.

tolerance

The deviation allowed from a given dimension.

top master key

The highest level master key in a master key system.

top of blade

The bitted edge of a single bitted key.

top pin

Usually a cylindrical shaped tumbler flat on both ends and installed directly under the spring in the pin stack.

total position progression

A process used to obtain key bittings in a master key system where bittings of change keys differ from those of the top master key in all bitting positions.

total stack height

Pin stack.

trim ring

Cylinder collar.

try-out key

a manipulation key which is usually part of a set, used for a specific series, keyway and/or brand of lock.

tubular key

A key with a tubular blade. The key cuts are made into the end of the blade, around its circumference.

tumbler

A movable obstruction of varying size and configuration in a lock or cylinder which makes direct contact with the key or another tumbler and prevents an incorrect key or torquing device from activating the lock or other mechanism.

tumbler spring

Any spring which acts directly on a tumbler.

two column progression

A process where key bittings are obtained by using the cut possibilities in two columns of the key bitting array.

two pin master key

A master key for all combinations obtained by progressing two bitting positions.

two step progression

A progression using a two increment difference between bittings of a given position.

U

unassociated change key

A change key which is not related directly to a particular master key through the use of certain constant cuts.

unassociated master key

A master key which does not have change keys related to its combination through the use of constant cuts.

uncoded

See uncombinated.

uncombinated

A cylinder which is, or is to be, supplied without keys, tumblers and springs.

uncontrolled cross keying

A condition in which two or more different keys under different higher level keys operate one cylinder by design. This condition severely limits the security of the cylinder and the maximum expan¬sion of the system, and often leads to key interchange.

unidirectional cylinder

A cylinder whose key can turn in only one direction from the key pull position, often not making a complete rotation.

universal keyway

Composite keyway.

vertical group master key

The two pin master key for all combinations listed in all blocks in a line down a page in the standard progression format.

V

VGM

Vertical group master key.

visual key control

Specification that all keys and the visible portion of the front of all lock cylinders be stamped with standard key code symbols.

VKC

Visual key control.

W

wafer

See "disc tumbler".

ward

Usually stationary obstruction in a lock or cylinder which prevents the entry and/or operation of an incorrect key.

ward cut

A modification of a key which allows it to bypass a ward.

wiggle key

See "manipulation key".

X

X

Symbol used in hardware schedules to indicate a cross keyed condition for a particular cylinder, e.g., XAA2, X1X (but not AX7).

Z

zero bitted

A cylinder which is or is to be combinated to keys cut to the manufacturer's reference number of bitting